Progress in Colloid & Polymer Science, Vol. 102 (1996)

PROGRESS IN COLLOID & POLYMER SCIENCE

Editors: F. Kremer (Leipzig) and G. Lagaly (Kiel)

Volume 102 (1996)

Gels

Guest Editor:

M. Zrínyi (Budapest)

Springer

SPRINGER-VERLAG BERLIN
HEIDELBERG GMBH

IV

ISBN 978-3-662-15604-9
ISSN 0340-255X

Die Deutsche Bibliothek –
CIP-Einheitsaufnahme

Gels : [proceedings of the Europhysics
Conference on Gels, held in
Balatonszeplak, Hungary, in September
1995] / guest ed.: M. Zrínyi.
   (Progress in colloid & polymer science ;
   Vol. 102)

NE: Zrínyi, Miklós [Hrsg.]; Europhysics
Conference on Gels ‹1995, Balaton-
széplak›; GT

© 1996 by Springer-Verlag Berlin Heidelberg
Originally published by Dr. Dietrich Steinkopff
Verlag GmbH & Co. KG, Darmstadt in 1996
Softcover reprint of the hardcover 1st edition 1996

Chemistry editor: Dr. Maria Magdalene
Nabbe; English editor: James C. Willis;
Production: Holger Frey, Bärbel Flauaus.

Type-Setting: Macmillan Ltd.,
Bangalore, India

ISBN 978-3-662-15604-9
DOI 10.1007/978-3-7985-1663-2

ISBN 978-3-7985-1663-2 (eBook)

Progr Colloid Polym Sci (1996) V
© Steinkopff Verlag 1996

Research and development of gels has experienced a rapid escalation, attracting worldwide interest and motivating the Macromolecular Board of the European Physical Society to organize a conference on gels.

The Europhysics Conference on Gels was held September 1995, in Balatonszeplak, Hungary. It was organized by the Macromolecular Board of the European Physical Society in cooperation with the Polymer Networks Group. The local organizers were the Hungarian Chemical Society and the Technical University of Budapest.

The purpose of this conference was to review the state of art and to present and discuss recent progress of gels and their applications and to look toward the future of gels. Since research into gels is a field which requires the development of multidisciplinary research collaborations, the major aim of the conference was therefore to bring together scientists from different disciplines and backgrounds in order to provide an excellent opportunity to exchange the latest scientific results and encourage further development. More than 160 participants from 26 countries attended the conference. The topics of contributions covered fundamentals and applications of gels formed from inorganic and organic polymers, colloidal particles and surfactant systems, new powerful methods such as scattering techniques, rheology, atomic force microscopy, swelling pressure and mechanical measurements, gels with sensitivity to changes in chemical and physical environment, novel technical and biomedical applications, computer simulation and new theoretical approaches.

The program was well-balanced between theoretical and practical aspects of gels and gelation. We gratefully acknowledge the financial support of the following institutions:

Commission of the European Communities Directorate General for Science, Research & Development, the Hungarian Academy of Sciences, the Hungarian Chemical Society, the Hungarian National Committee for Technological Development, Technical University of Budapest, IFHERD and FEFA Foundations.

Their support has made it possible to invite scientists from countries with limited economic resources, making the meeting a truly international event.

This Progress Volume contains a selection of the papers presented at the Europhysics Conference on Gels. It is hoped that it demonstrates the manifold nature and diversity of gels as an interdisciplinary science.

Miklós Zrínyi

Progr Colloid Polym Sci (1996) VII
© Steinkopff Verlag 1996

CONTENTS

VIII

Progr Colloid Polym Sci (1996) 102:1–3
© Steinkopff Verlag 1996

H. Galina
J. Lechowicz

Monte–Carlo modeling
of polymer network formation

Dr. H. Galina (✉) · J. Lechowicz
Rzeszów University of Technology
Faculty of Chemistry
W. Pola 2
35-959 Rzeszów, Poland

Abstract Monte–Carlo simulation of a step-wise homopolymerization of 3-functional monomer has been performed. The bonds were formed irreversibly between units selected at random (mean-field approximation). The time correlations were introduced by imposing substitution effects upon reactivity of functional groups. The cycle formation was modeled assuming Gaussian behavior of all bonds introduced in simulation. The probability of intramolecular reaction was granted higher values compared to those for intermolecular ones by applying single cyclization parameter. Quite surprisingly, the effect of cyclization parameter on the gel point conversion turned out to become an S-shaped curve.

Key words Monte–Carlo simulation – aggregation – cyclization – gel point – model polymerization

Introduction

Any model of polymer formation is about the build-up of connectivity between units in the system. Usually, one starts with a set of units and looks at a chemical or physical process of introducing bonds between pairs of units. The problem is *at which moment* and *between which pair of units* (or functional groups) the bond is formed.

In the models with mean-field approximation, the spatial distribution of units is disregarded. This means that any effects due to slow diffusion of reactants or concentration fluctuations (e.g., at very early stages of vinyl-divinyl copolymerization [1, 2]) cannot be adequately taken into account. The same applies to cyclization reactions. Prior to gel point the mean concentration of units (groups) belonging to the same molecule is zero in the limit of a macroscopic system. Hence, the probability or rate of such a reaction is automatically set to zero irrespectively of dilution of the system.

Among the mean-field models there are two limiting cases with respect to how they deal with time correlations, i.e., the moments when two units (groups) meet to form a bond. In the statistical models (including the fundamental Flory–Stockmayer [3, 4] one) the units (groups) bear no memory on the extent of reaction (or time) at which they happen to react. Note that only for the systems in equilibrium is such a memory irrelevant. For the systems where bonds are formed in irreversible reactions, kinetic models seem to be more adequate [5–7]. Unfortunately, the kinetic models applied to modeling of network formation fail beyond the gel point.

On the other hand, there are percolation models which constitute a class of Monte–Carlo methods dealing with connectivity build-up. Spatial and time correlations including cycle formation are built in naturally in the models. The percolation models, however, both lattice or off-lattice ones, suffer from inadequate mobility of reacting units. Besides perhaps the most sophisticated off-lattice algorithms [8, 9] their applicability in modeling the real systems seems to be limited to the vicinity of gel point.

Dušek et al. [10, 11] have devised a simple Monte–Carlo algorithm for studying homopolymerization of a 3-functional monomer with the mean-field assumptions and allowing for time correlations and random cyclization reactions to be taken into account.

We have developed a mean-field Monte–Carlo model of the same class as that of Dušek et al. in order to study systems with high tendency to cyclization. The same step-wise homopolymerization of a 3-functional monomer was studied. By changing a single parameter we have forced intramolecular reactions to proceed with higher priority than intermolecular ones. The magnitude of the parameter can roughly be referred to as degree of dilution of the system. Below we present preliminary results of application of the algorithm.

The model

Details of the algorithm have been described elsewhere [12]. The units to react were selected at random in the whole system (mean-field approximation). The functional groups reacted with the first shell substitution effect [13, 14]. In short, the reactivity of all three functional groups at the monomer was the same and set to $k_0 = 1$. As soon as one of them had reacted, the reactivity of the remaining two changed to k_1. The same applied to the reactivity k_2 of the only group that remained on doubly reacted unit. Thus, the rate of reaction between the groups on units of substitution degrees i and j was the product of respective reactivities pre multiplied by numbers of functional groups available to reaction: $(3 - i)(3 - j)k_ik_j$. If the selected units happened to belong to the same molecule, the rate was $Cl^{-3/2}(3 - i)(3 - j)k_ik_j$ where C is the arbitrary cyclization parameter, and l the shortest number of bonds between the selected units. The form $l^{-3/2}$ suggests that every bond at the path linking units is a Gaussian sub chain.

Results

The simulation was carried out for systems of different size. The upper limit of the size was determined by the available computer memory. The critical conversion at the gel point, p_c was determined as the highest slope in the plot $\log P_w$ vs. conversion (P_w is the weight-average degree of polymerization in the system). In our simulation the accuracy of this method of gel point determination depended on the system size as illustrated in Fig. 1 where the standard deviation of p_c determination is plotted against the system size. Each critical conversion was determined in at least seven simulation runs performed for the same input parameters. Note that it was standard deviation, not the average gel point conversion that depended on the system size. The system size used in further experiments was 10^5 units. The cyclization parameter was changed in broad range of 0 through 1500.

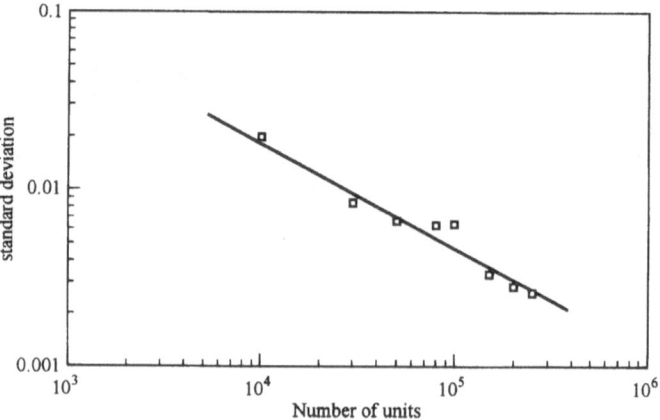

Fig. 1 Standard deviation of gel point determination vs. system size (number of units). The average gel point conversion does not depend on the size

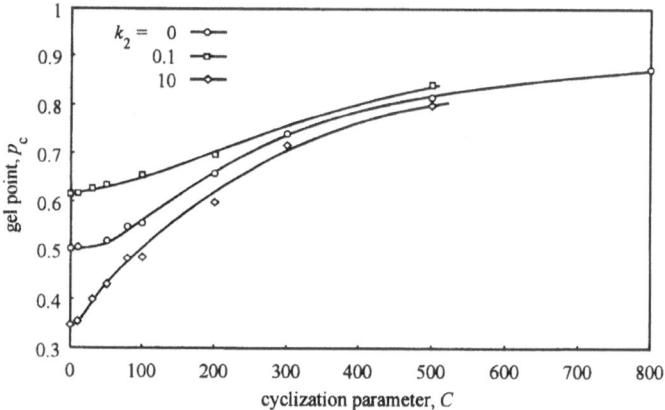

Fig. 2 The shift in gel point conversion vs. cyclization parameter. The rate constant k_2 indicate the reactivity of the third functional groups in di-substituted unit relative to that in monomer or mono-substituted unit. Note the S-shape of curves

Selected conversions at the gel point for three systems differing in substitution effects are presented in Fig. 2. For each curve both k_0 and k_1 were set to 1. To our surprise the shift in the gel point with the magnitude of cyclization parameter C was not uniform, but the curves were slightly S-shaped. The effect is best seen for the "random" system, with no substitution effects imposed on the reactivity of functional groups. Other authors have not observed, anything similar in their Monte–Carlo [15] or analytical [16] studies of cyclization.

We have found the relatively rapid shift in gel point conversion at a certain range of C difficult to explain. One possible explanation might be the extent of cyclization prior to gel point. For relatively small value of C, our system, which is based on the mean-field approximation, contains negligible number of cycles until well beyond the gel point. The rapid shift in the gel point conversion

Progr Colloid Polym Sci (1996) 102:1–3
© Steinkopff Verlag 1996

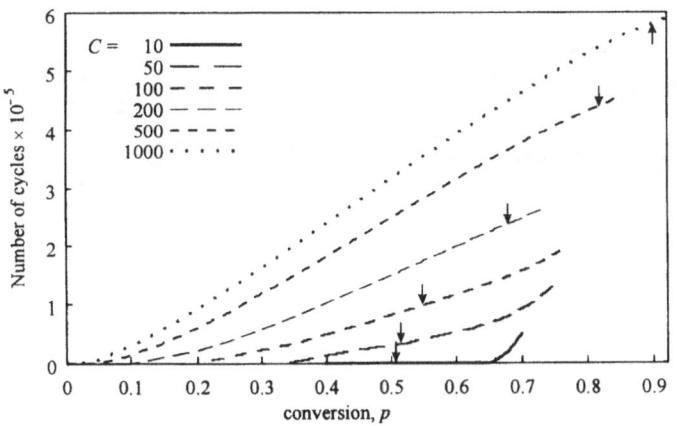

Fig. 3 The total number of cycles in polymer species (including gel molecule) for different cyclization parameters. The arrows point at the gel point

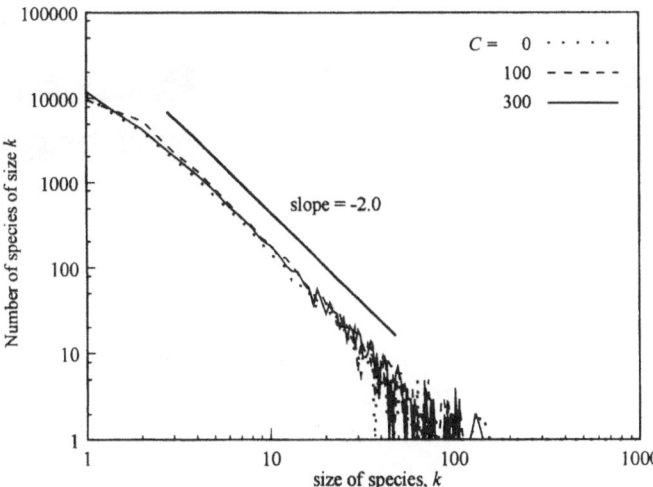

Fig. 4 The size distribution of polymer species at the gel point for systems with and without forced cyclization

roughly corresponds to the situation when a substantial number of cycles appears prior to gelation. This is illustrated in Fig. 3.

Unfortunately, while analyzing critical behavior [17] of our model, we were unable to determine the critical exponent γ in the relation $P_w \sim |p_c - p|^{-\gamma}$. Probably, the system containing 10^5 units was too small to observe either classical or percolation value of γ. Similarly, we were not able to determine the exponent in the relationship $w_g \sim |p_c - p|^\beta$ (w_g is the content of gel fraction).

According to van Dongen and Ernst [18], at the gel point the scaling relation for the concentration of cluster size, c_k is

$$c_k(p_c) \sim k^{-\tau} \qquad (1)$$

with $\tau = 5/2$. As can be seen in Fig. 4, in our model the value of τ is somewhat smaller and equals 2. It does not depend on cyclization parameter C.

In conclusion, one can write that the algorithm devised for studying polymer network formation models reasonably well the process of network formation. Except for the S-shaped curves in Fig. 2 all results were predictable. The system size was large enough to study the shift in the gel point conversion for systems with forced cyclization and time correlations. The critical behavior of these gelling systems has to be studied using more powerful computers.

References

1. Dušek K, Galina H, Mikeš J (1980) Polym Bull 3:19–25
2. Okay O, Kurz M, Lutz K, Funke W (1995) Macromolecules 28:2728–2737
3. Flory JP (1953) Principles of Polymer Chemistry. Cornell Univ Press, Ithaca
4. Stockmayer WH (1942) J Chem Phys 11:45–55
5. Kuchanov SI (1978) Methods of Kinetic Calculations in Polymer Chemistry (in Russian) Izd Khimia, Moscow
6. Galina H, Szustalewicz A (1990) Macromolecules 23:3833–3838
7. Faliagas AC (1993) Macromolecules 26:3838–3845
8. Leung YK, Eichinger BE (1984) J Chem Phys 80:3877–3884
9. Lee KJ, Eichinger BE (1989) Macromolecules 22:1441–1448
10. Mikeš J, Dušek K (1982) Macromolecules 15:93–99
11. Šomvarský J, Dušek K (1994) Polym Bull 33:369–376
12. Galina H, Lechowicz J (1995) Comp Polym Sci 5:197–201
13. Gordon M, Scantlebury GR (1966) Proc Roy Soc (London) A292:604–621
14. Dušek K, Prins W (1969) Adv Polym Sci 6:1–102
15. Stepto RFT (1995) Private communication
16. Faliagas AC (1994) Private communication
17. Stauffer D, Coniglio A, Adam M (1982) Adv Polym Sci 44:103–157
18. van Dongen PGJ, Ernst MH (1987) J Stat Phys 49:879–926

Progr Colloid Polym Sci (1996) 102:4–8
© Steinkopff Verlag 1996

V.I. Irzhak
S.E. Varyukhin
T.F. Irzhak

Relaxation properties of polymer gels and concept of physical networks

Dr. V.I. Irzhak (✉) · S.E. Varyukhin
T.F. Irzhak
Department of Polymer and
Composite Materials
Institute of Chemical Physics
in Chernogolovka
Russian Academy of Sciences
Chernogolovka
Moscow distr., 142432, Russia

Abstract Relaxation properties of
linear and network polymers are
analyzed on the basis of physical
networks.

Two approaches are proposed:
a) taking into account the equilibrium
concentration of physical crosslinks;
b) taking into account the concentration of nonequilibrium physical
crosslinks reflecting the nonequilibrium chain conformation and
kinetics of their transformation to
equilibrium ones. The modulus
changing with time and/or temperature is shown to connect with
integral function of molar mass
distribution of polymer chains.

Key words Physical networks –
relaxation – thermomechanics

Introduction

The main idea on which theoretical description of
the polymer relaxation properties at temperatures above
T_g is based consists of a tube assumption: the coefficients
of translation diffusion along and across the chains are
rather different [1]. This mechanism is conditioned by the
existence of topological restrictions, i.e., the macromolecule is supposed to be inside a tube with walls formed
by other macromolecules. Although many aspects of the
relaxation behavior of polymers were described on that
basis, there are experimental results that did not coincide
quantitatively with the theoretical forecasts [2]. The
probable cause of this is the fact that the relaxation mechanism of the tube itself is not clear enough. The description of this process is rather difficult due to that its
mechanism must take into account macromolecular interactions and more complicated movement occurring in the
system.

The consideration of a wide range of properties of
polymer systems has been based on the concept of physical
networks [3–5]. The latter allows to clearly imagine rather
complicated intermolecular interactions. At the same time,
such approach allows to quantitatively describe many
phenomena.

Model of physical networks is used for polymer gels
particularly. Many of their properties are caused by existence of physical knots of different nature [6]: microphase
regions, clusters, i.e., regions of high order and density
(glass nuclei), strong bonds (hydrogenous, ionic and so on).
Thus, it is necessary to considerate gel properties from the
point of view of physical networks.

Here, some relaxation properties of linear and network
polymers are considered based on the model of physical
networks.

Model of physical networks

The main difference of physical networks from chemical
ones is that their crosslinks have limited lifetime depending
on temperature and mechanical effects. In [3] the relation
for equilibrium concentration of crosslinks (n) was obtained:

$$n = \frac{D \exp(D) - \exp(D) + 1}{(\exp(D) - 1)^2}.$$

(1)

Progr Colloid Polym Sci (1996) 102:4–8
© Steinkopff Verlag 1996

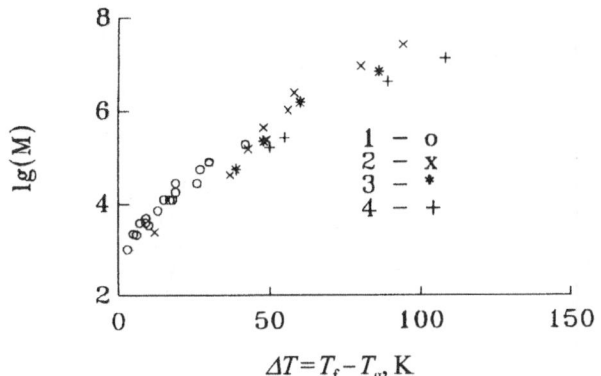

Fig. 1 Correlation dependence of molar mass and fluidity temperature for different polymers. Symbols ○, ×, *, and + relate to polyurethanes, polystyrenes and polyacrylonitrile, accordingly

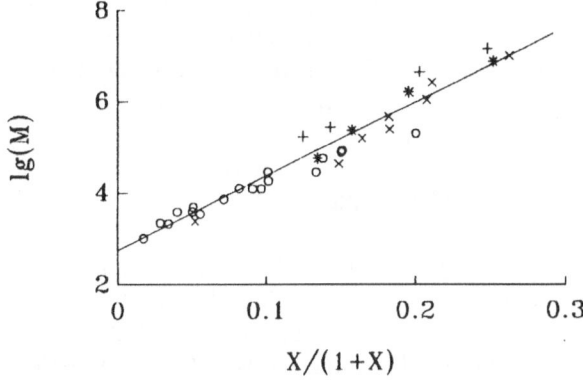

Fig. 2 The same relation in coordinates of Eq. (5)

Here, $D = ((E - TS - f(\lambda))/RT)$, $f(\lambda)$ is a function expressing the stress effect on free energy of crosslinks, λ is strain of the chains joining to the crosslink, E and S are energy and entropy of the crosslink formation, T is temperature. Equation (1) allows to relate flow temperature of linear polymers with their molar mass basing on concept of physical networks [7].

Physically, flow temperature T_f is considered as gel point, i.e., when crosslinks concentration is equal approximately to chain concentration.

$$n/N \cong M \frac{D \exp(D) - \exp(D) + 1}{(\exp(D) - 1)^2} = 1 \qquad (2)$$

where N is chain number, M is their molar mass.

In Fig. 1 are data showing the correlation between T_f and molar mass of linear polymers with narrow molar mass distribution (data of [7]). One can see the uniform relation is realized for any polymer if $\Delta T = T_f - T_g$ is used. In addition, temperature beginning of high elastic plateau corresponds to values of kinetic segments: $M \cong 10^3$. This means that high density fluctuations take an important role in the physical network formation. Probably, they play a role in glass nuclei [8] with low mobility of segments and high order of structural organization [9].

These results can be explained on the basis of Eq. (2).

Actually, if at T_g molar mass of chains of physical network is supposed to be equal to Kun's segment M_g and at T_f crosslinking index is equal to 1, from (2) we can obtain:

$$\ln(M/M_g) \cong D - D_g = (E - f(\lambda)) \frac{T_f - T_g}{R T_f T_g}. \qquad (3)$$

As one can see, (3) has the form of equation WLF and expresses the uniform relation between M and ΔT:

$$T_f = T_g + \Delta T = T_g(1 + X), \qquad (4)$$

where $X = \Delta T/T_g$.

And then:

$$\ln(M/M_g) = \frac{E - f(\lambda)}{R T_g} \frac{X}{1 + X} = \frac{AX}{1 + X}. \qquad (5)$$

Equation (5) corresponds well to experimental data (Fig. 2). This is possible if value A is uniform for any polymer system. Actually, the uniformity of A was shown in [10].

$$A = \frac{E}{R T_g} \cong 26.$$

In [7] there was analysis of correlation of thermomechanical curves with MMD of linear polymers. It was shown that the temperature dependence of relaxation modulus reflects the integral function of MMD.

$$\frac{\sigma}{E\varepsilon} = 1 - \int_0^s dM \omega(M) \int_{-\infty}^{T_f T(M)} \delta(t)\, dt. \qquad (6)$$

Here, σ, E and ε are stress, modulus and deformation, accordingly, M and $\omega(M)$ are molar mass and value of its weight fraction, T and T_f are experimental temperature and flow temperature corresponding to the molar mass M; $\delta(t)$ is impulse function: it is equal to 0 for all $t \neq 0$, $\delta(0) = 1$; so, the second integral is equal to 0 at $T < T_f$ and to 1 at $T \geq T_f$. The right part of the equation is the integral function of MMD. Therefore, Eq. (6) expresses the connection of experimental parameters of the high elastic plateau (σ, E and ε) with integral function of MMD.

There are experimental data in [7] supporting this approach.

Note, Eq. (1) does not contain any time term because we deal with quasi-equilibrium state. But it is well known that the parameters of high elastic plato have relaxation character, i.e., dependent on time. Therefore, it is necessary to consider in more detail the state of a polymer chain in the system under deformation.

Relaxation of physical networks

Any physical crosslink has limited lifetime. Therefore, momentary loaded stress leads to affine deformation of the physical network, i.e., distances between crosslinks change in proportion to external dimensions of the body. Before the stress loading interknot chains are supposed to be in equilibrium state: this means the chains' dimensions (distances between topologically neighbor network knots) and their molar mass are connected by definite relation (in case of Gaussian chains $\langle r^2 \rangle = nl^2$). After the stress loading this relation should be disturbed: interknot chains will be in nonequilibrium conformation state. In other words, the corresponding knots will be in nonequilibrium.

The relaxation process is proposed to consider the recovery of equilibrium chain conformation or, that is, the transformation of nonequilibrium knots to equilibrium ones. So, the relaxation process can be expressed either as the diffusion of polymer chains or as the transformation of nonequilibrium knots of physical network to equilibrium ones.

Generally, the diffusion of macromolecules in condensed system is considered as a Rousen process inside the tube followed by changing of the tube configuration [11, 12]. The latter process is supposed to obey Rousen law with essentially higher relaxation time.

Transformation of nonequilibrium knots of physical network is proposed to be an alternative approach to describe relaxation process, moreover, the use of the simplest kinetic law is supposed to be possible:

$$dx/dt = - x/\tau . \tag{7}$$

Here, x is concentration of nonequilibrium knots, τ is the relaxation time.

It is obvious that the relaxation time τ is not constant in the course of the process. Besides dependence upon the system state parameters (temperature, pressure, polymer concentration) and the chain molar mass M, τ must be dependent on x. Actually, if a chain is characterized by lot of nonequilibrium knots the probability of transformation of one of them is very low because of nonequilibrium conformation of the whole chain. This probability must be greater the closer the conformation to equilibrium state. Consequently, it is reasonable to suppose τ is the function of x. The power type of this relation is the simplest.

$$\tau = \tau_0 x^b , \tag{8}$$

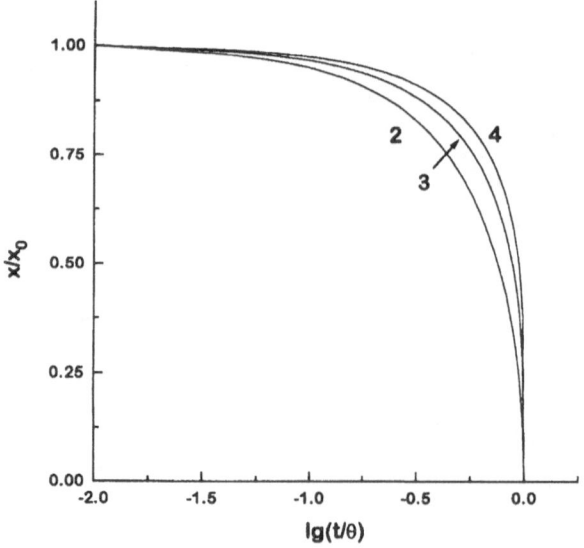

Fig. 3 Curves of Eq. (9). There are values of "b" near the curves

where $b > 1$.

Taking into account (8) from (7), we have:

$$x = x_0 (1 - t/\theta)^{1/b}, \tag{9}$$

where $\theta = x_0^b \tau_0 / b . \tag{10}$

x versus t is shown in Fig. 3 for several values of b. As one can see, curve $x(t)$ decays to 0 in narrow time interval. Moreover, the larger b, the narrower this interval. Therefore, stepwise function $u(t)$ is supposed as reasonable approximation of (9):

$$u(t) = \begin{cases} 1 & \text{at } t < \theta, \\ 0 & \text{at } t \geq \theta. \end{cases} \tag{11}$$

For description of deformation process in the simplest case, we can use the Kelvin model:

$$d\varepsilon/dt = (\sigma - E(x)\varepsilon)/\eta . \tag{12}$$

Here, ε is deformation, σ is strain, $E(x)$ is modulus, and η is viscosity.

Stress relaxation or creep are connected with concentration x via $E(x)$. In the system of linear polymers $E(x) \neq 0$ if $x \neq 0$, i.e., there are nonequilibrium knots. When $x = 0$ $E(0) = 0$, Eq. (12) describes the polymer flow:

$$d\varepsilon/dt = \sigma/\eta . \tag{13}$$

In the first approximation the use of the classical theory of high elasticity is possible to express connection of modulus with the crosslinks' concentration:

$$E(x) \cong kTfx/2 = kTfx_0 u(t) = E_0 u(t) , \tag{14}$$

where f is the crosslink functionality.

Progr Colloid Polym Sci (1996) 102:4–8
© Steinkopff Verlag 1996

While $t < \theta$, Eq. (12) describes the normal law of creep or stress relaxation. At the steady state:

$$\varepsilon = \sigma/E_0 . \tag{15}$$

Note, as far as E_0 depends on x_0 and the latter is in proportion to molar mass of interknot chain at T_g (M_g), the values of modulus and deformation are affected by value of kinetic segment at T_g. In other words, molar mass of the chain between entanglements (M_e) is the function of M_g. This phenomenon was found in experiments [13], but there has been no physical explanation to date. A model of physical networks allows us to understands this.

When time reaches the moment θ, $E(x)$ will be equal to 0 and the steady state will terminate. Remember that $\theta = x_0^b \tau_0/b$.

This relation gives the connection between molar mass, temperature and other parameters of the system state at beginning of the polymer flow. Unlike Eq. (5), there is the time parameter θ in this relation. Thus, Eq. (10) expresses the principle of temperature-time equivalence and θ is equivalent to the shift factor a_T.

There are data dealing with monodisperse polymers [14, 15] that have shown that decay of the curve of dependence of relaxation modulus on time (frequency) are connected with molar mass according to Eq. (10).

So, using the model of physical networks it is possible to describe relaxation processes.

In case of polydisperse polymer, the relation for relaxation modulus is:

$$E \cong \sum n_i x_i = \sum n_i x_{0i} u_i(t) , \tag{16}$$

where index i relates to polymer fraction with molar mass M_i.

As far as x_{0i} is proportional to M_i, product $n_i x_{0i}$ is in proportion to weight fraction of polymer chains with molar mass M_i. Note, the dependence of total modulus on weight fractions of the polymer is obtained naturally, without any suppositions.

At the beginning stages of the process while $t < \theta_{min}$, i.e., all chains do not relax and all fractions are included in the sum of Eq. (16) with coefficients equal to 1. In the course of the process the smallest chains will be relaxed and acquire the equilibrium conformation: for them $t > \theta_i$. Now, these fractions take the coefficients equal to 0. The value of the modulus decreases in proportion to their weight fractions. Therefore, time law of the modulus changing in the relaxation process reflects integral function of polymer MMD.

In [16, 17] bifraction polymers were studied. It was shown that the modulus decreased at that time (frequency) which corresponds to end of high elastic plato of the fraction of lower molar mass. The value of decrease was in proportion to weight fraction of latter. These results support the validity of Eq. (16).

It makes no difference whether these polymer systems are linear or crosslinked. The process of chain relaxation obeys the same law: time (temperature) of termination of relaxation process depends on fraction of molar mass in a similar manner. The difference consists in the following: termination of relaxation process leads to flow in case of linear polymers and to equilibrium modulus in case of network ones.

Many examples of connection of MMD of network chains with value of deformation (relaxation modulus) in the temperature region from T_g to equilibrium state were presented in [18].

Conclusion

As one can see, the approach based on the model of physical network is very useful, although phenomenological. The task is to clarify the physical sense of the model. Equation (8) is proposed to express the dependence of relaxation rate on chain conformation, i.e., on the system nonequilibrium degree. We believe the process of the nonequilibrium knots transformation is analogous to tube changing. We believe the methods of physical kinetics and/or computer simulation will allow to obtain a more correct version of Eqs. (7) and (8).

References

1. De Genne P-G (1979) Scaling Concept in Polymer Physics. Cornell Univ Press; Ithaca, NY
2. Rubinstein M, Colby RH (1988) J Chem Phys 89:5291–5306
3. Solovyov MYe, Raukhvarger AB, Irzhak VI (1986) Vysokomol Soed 28B: 106–110
4. Raukhvarger AB, Solovyov MYe, Irzhak VI (1989) Chem Phys Lett 155: 455–459
5. Solovyov MYe, Raukhvarger AB, Ivashkovskaya TK, Irzhak VI (1992) In: Progress in Coll Polym Sci 90:174–177
6. Keller A (1995) In:
7. Olkhov YuA, Baturin SM, Irzhak VI (1996) Vysokomol Soed 38A:549–554
8. Rostiashvili VG, Irzhak VI, Rozenberg BA (1987) Steklovaniye Polimerov (Glass Transition in Polymers) Khimiya, Moskva

9. Sanditov DS, Kozlov GV, Belousov VM, Lipatov YuS (1992) Ukranian Polymer Journ 1:241–258

10. Matveev YuI, Askadskii AA (1991) Vysokomol Soed 33A:1251–1255

11. Doi M, Edvards SF (1986) The Theory of Polymer Dynamics. Clarendon, Oxford

12. Mark JE, Eisenberg A, Graessley WW, Mandelkern L, Koenig JL (1984) Physical Properties of Polymers. Am Chem Soc, Washington DC

13. Van Krevelen DW (1972) Properties of Polymers Correlations with Chemical Structure. Elsevier Publ, Amsterdam-London-NY

14. Fujimoto T, Dzaki M, Nagasawa M (1968) Journ Polymer Sci A2 6:129–137

15. Vinogradov GV, Dzyura EA, Malkin AYa, Grechanovskii VA (1971) Journ Polymer Sci A2 9:1153–1167

16. Onogi S, Masuda T (1968) Kobunsi 17:640 In: Ferry JD (1968) Viscoelastic Properties of Polymers, 2nd ed, John Wiley & Sons NY

17. Seidel U, Stadler R, Fuller GG (1994) Macromolecules 27:2066–2072

18. Irzhak TF, Varukhin SE, Olkhov YuA, Baturin SM, Irzhak VI (1996) Vysokomol Soed (in press)

Progr Colloid Polym Sci (1996) 102:9–14
© Steinkopff Verlag 1996

F.A. de Wolf
R.C.A. Keller

Characterization of helical structures in gelatin networks and model polypeptides by circular dichroism

Dr. F.A. de Wolf (✉)
Agrotechnological Research Institute
(ATO-DLO)
P.O. Box 17
6700 AA Wageningen, The Netherlands

R.C.A. Keller
Department of Biochemistry of Membranes
Centre for Biomembranes and
Lipid Enzymology
University of Utrecht
Padualaan 8
3584 CH Utrecht, The Netherlands

Abstract Reversible physical crosslinks (junction zones) in gelatin gels and films are thought to consist of triple helical collagen-like structures, however other structures may also occur. In an attempt to understand the relation between the material properties and the various junction structures and their abundance, we characterized these structures using circular dichroism (CD) as a function of the conditions applied. It appeared that the commonly used reference CD spectra for random coil and 3_1 helix could not be used to deconvolute the spectra observed for gelatin. Most probably, the commonly-used random coil reference spectra do not really reflect unordered structure, while the random coil reference based on the cationic form of polyLysine actually reflects a clear helicity. Denatured gelatin or collagen were found to be better approximations of an unordered polypeptide structure. The prolyl chromophore was shown to have a pronounced influence on the shape and position of the CD spectra. Thus, it was concluded that only polypeptides with the appropriate prolyl content can be used as references, while analysing the relative secondary structure contents by CD. By using temperature-dependent CD, we were able to discern multiple and single helical structures at high and low gelatin concentrations, respectively, by the observed cooperativity of helix-to-coil transition.

Key words Gelatin – circular dichroism – proline content – random coil – thermal denaturation – model peptide

Introduction

Gelatins are prepared from alkaline- or acid-solubilized (i.e. chemically modified) collagen type I or III by extracting the protein from hides or decalcified bones under denaturing conditions (i.e. at elevated temperature).

The defining structural element of all collagens is the right-handed triple helix, which results from 1) a high imino acid content, 2) inter-chain hydrogen bonding, and 3) a repetitive sequence of glycine-X-Y amino acid triplets. The precise location of the glycine is essential, since its small size fits into the central part of the triple helix. This picture has been deduced mainly from x-ray diffraction studies on collagen fibres (e.g. refs. [1, 2]) and collagen-like model peptides [3]. Probably as a result of the high imino acid content, each of the three individual polypeptide chains in the triple helix forms an extended left-handed helix closely similar to the polyProline type II conformation.

At elevated temperatures, the triple helix is denatured, the individual gelatin chains lose their left-handed helical conformation and assume a more random structure [4, 5]. In this situation, a slow cis-trans isomerization of the proline-preceding peptide bonds occurs [6, 7]. Typically, some 10–15% of these bonds are in the cis conformation in gelatin [8].

When solutions of gelatin are cooled, the polypeptide chains usually undergo a frustrated renaturation process, in which part of the chains again assume a helical conformation of the polyProline-II type and part of the chains intertwine again to form multiple helices [4, 9–13]. Many of the multiple helical parts function as physical crosslinks (junction zones) that give rise to gelation [14–18]. The renaturation is slowed down by the slow isomerisation of the cis peptide bonds [19], and multiple helix formation is also hindered by steric effects, chain entanglements and topological restrictions resulting from chains taking part in more than one junction zone.

Although it is generally thought that in gelatin systems, helix stabilization is exclusively due to intermolecular triple-chain associations, other ways of helix stabilization may also be possible (e.g. [4, 5, 9, 13, 20]).

In synthetic polypeptides, stable extended left-handed helices of the type occurring in collagen can occur also in single chains [22–24]. The stabilization can be of electrostatic nature (charge repulsion in poly-ions, see ref. [24]) or of steric nature, in combination with a relatively small entropy change associated with the coil-helix transition (in polypeptides with a high imino acid content, see ref. [25]). In water, polyProline assumes a relatively "open" type II conformation that closely resembles the left-handed helices in collagen, and in which the peptide bonds are in the trans conformation (97–98% trans) [26, 27]. In contrast to poly(Prolyl-Glycyl-Proline) [28], polyProline cannot form triple helices due to the absence of glycines. The single polyProline II helices are relatively stable: they are not denatured at high temperatures in water [29], but only in the presence of high concentrations of (chaotropic) calcium ions [30]. PolyLysine and several other charged polypeptides apparently form single helices that can be denatured by temperature [29, 31–33] and chaotropic ions (Ca^{2+}, Li^+) [30, 31].

To date, it is not known whether, upon cooling of gelatin, the chain sections with high helix-forming propensity [34–36] can form single helices, for example at high gelatin concentration, where topological effects may hinder the formation of multiple helices, or at extremely low gelatin concentration, i.e. in the range where the renaturation kinetics are first order with respect to gelatin concentration [9]. The melting behaviour of poly(Glycyl-Prolyl-Alanine), which is a model for gelatin, was show to be non-cooperative under some conditions [37]. This is characteristic for single helices [29, 31, 33].

The secondary structure elements in protein solutions and gels can in principle be determined by circular dichroism (CD), although the deconvolution of gelatin spectra into defined secondary structure elements appears to be somewhat problematic. The well-established method of Greenfield and Fasman [38] has been successfully employed in studies of polypeptides with high contents of α-helix or β-sheet (see for example refs. [39–41]). However, the interpretation of the spectra of proteins and polypeptides with a high content of random structure is still not satisfactory (see for example ref. [41]). This can be partly explained by the intrinsic difficulty of defining a "random" polypeptide structure, but the deconvolution problems are probably also related to the reference spectrum based on cationic polyLysine, which has the features of an extended polyProline II helix rather than random structure [22–24].

In view of the lack of reliable reference spectra for the gelatin system, we characterised and attempted to define in the present work the CD spectra of different secondary structure elements that are potentially relevant as reference spectra for the gelatin system.

Experimental

Materials

Gelatin was obtained from Gelatine Delft B.V., Delft, the Netherlands (batch 7725). Type I collagen (Sigma "type III") and polyamino acids were purchased from Sigma, Bornem, Belgium, except for poly(Gly-Pro-Ala), which was from Bachem, Bubendorf, Switzerland. The prePhoE signal peptide (20 residues) and apocytochrome c (104 residues) were obtained and treated as described before [42, 43]. The approximate average lengths of the poly amino acids were: polyLysine: 234–547, poly(Pro, Thr, Arg): 42–169, poly(Gly-Pro-Ala): 180, poly(Pro-Gly-Pro): 63, and polyProline: 104 amino acids. Sodium caseinate (a mixture of approximately 43 mol% $α_{s1}$-, 11% $α_{s2}$-, 32% β- and 14% κ-casein from cow milk, with 199, 207, 209 and 169 residues respectively) was obtained from DMV international, Veghel, the Netherlands. Other chemicals were of analytical grade or better.

Preparation and analysis of the samples

The water- and nitrogen content of the protein preparations were determined by weight (before and after drying, using P_2O_5 or a vacuum oven) and by Kjeldahl or element-analysis (Kjeltec Auto 1030 analyzer, and Carlo Erba Instruments CHNS-O AE 1108, respectively). Thus, we found the "nitrogen weight-to-gelatin weight" conversion factor to be 5.40, in accordance with refs. [44, 45]. Solutions were made with milli-Q water.

All circular dichroism measurements were performed on a Jasco J-600 spectropolarimeter essentially as described

Progr Colloid Polym Sci (1996) 102:9–14
© Steinkopff Verlag 1996

before [40]. Quantitative deconvolution of CD-spectra in terms of predefined reference spectra was performed by computer, according to the linear least squares method (see for example ref. [46]).

Results and discussion

Deconvolution of CD spectra

Figure 1 shows the CD spectra of gelatin above and below the melting temperature. The spectra clearly monitor the difference in the conformation of the gelatin molecules at the two temperatures. In order to understand the molecular behaviour of gelatin, and in order to be able to correlate this behaviour with the rheological (i.e. macroscopical) behaviour of the system, a quantitative interpretation of the spectra in terms of defined secondary structure elements is required. Depending on the temperature, we expected a contribution from triple helical, unordered and possibly also single helical elements. Therefore, we initially used the widely-applied references for these structures: collagen, positively charged polyLysine and polyProline in the type II (3_1 helix) conformation. (The latter conformation is spontaneously assumed by proline in water, and is stable even at high temperatures). The CD spectra of these references are shown in Fig. 2. We also paid attention to a possible contribution of *cis* peptide bonds, by using the polyProline I spectrum as reference (i.e. spectrum of the all-*cis* form of polyProline) (e.g. ref. [27]). However, we never observed a significant contribution.

It appeared that the set of references shown in Fig. 2 could not be applied to gelatin. For example, this would lead to the conclusion that an increase of "unordered" structure could be induced by cooling the gelatin solution below the helix-coil transition temperature, which is obviously wrong. This appears also from Fig. 2, which shows that the polyLysine spectrum, commonly used as "random coil" reference (see for an overview ref. [47]), resembles the spectrum of triple helical collagen.

Using sets of globular reference proteins with known crystal structure, virtual CD spectra of unordered structure elements have also been algebraically derived from the CD spectra of these proteins as the common spectral elements that cannot be ascribed to α-helical or β-structure (see for example [47]). We found that, like polyLysine, these references for "unordered" structure could not be applied to materials with a low content of α-helix or β-structure, such as gelatin or casein (data not shown). Probably, the residual non-α helical/non-β domains in globular proteins are not truly "unordered", i.e. their local conformation cannot be considered as random samples

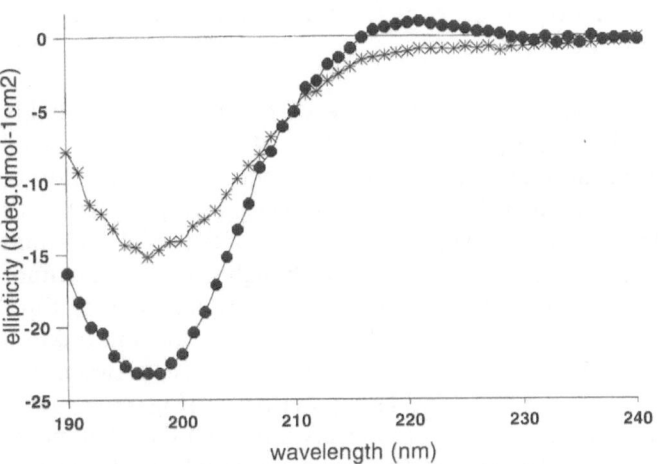

Fig. 1 Circular dichroism (CD) spectra of 0.05% (w/v) gelatin in a partially denatured state at 40 °C (∗), and in a partially renatured state after 70 h at 4 °C (●), both in 10 mM NaCl, 15 mM EDTA, 100 mM Na-phosphate buffer pH 7.0

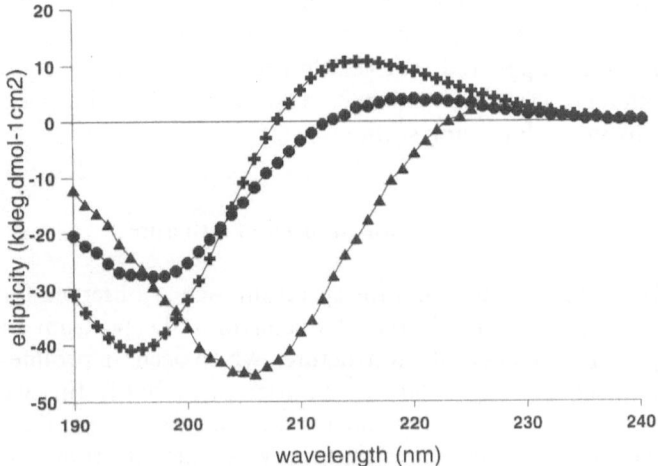

Fig. 2 CD spectra of charged polyLysine (+), "native" collagen (●) and polyProline (▲) as potential references for random coil, triple helix and single 3_1 helix, respectively. The spectrum of polyLysine (0.01% (w/v), pH 5.7, 22 °C) was taken from refs. [38, 47]; the spectrum of collagen (0.0007%, w/v) was recorded in 1 mM NaCl, 1 mM Na-phosphate, 0.15 mM EDTA at 4 °C; that of polyProline (0.05%, w/v) in water at 4 °C

of the total conformational space that is energetically allowed for polypeptides.

We now believe that gelatin or collagen, "freshly" but thoroughly denatured at temperatures well above the helix-coil transition, is a much better reference for unordered gelatin structure than positively charged polyLysine or the residual non-α/non-β domains in globular proteins. However, before addressing this point, we will first consider an explanation for the negative results described above.

Evaluation of reference spectra – spectral influence of prolyl residues

As mentioned, the structure of polyLysine is probably predominantly single helical [22–24]. It can be denatured to an unordered structure only under certain conditions, for example at high temperature (e.g. refs. [29, 30]). This accounts for the apparent similarity of the polyLysine and collagen spectra (Fig. 2), even though polyLysine cannot form triple helices, which require a glycine in every third position. If polyLysine assumes a 3_1 helical conformation like polyProline, the observed difference between the spectra of polyLysine and polyProline (Fig. 2) is rather unexpected. The data in Fig. 3A show that the difference is caused by the effect of the prolyl chromophores on the CD spectrum: with increasing imino acid content, the spectra were increasingly red-shifted. We observed that, in addition to a red shift, the presence of prolyl residues also changes the form of the spectra (Fig. 3B).

The present analysis shows that the imino acid content of reference compounds for secondary structure elements should closely correspond to the imino acid content of the proteins to be analyzed in terms of these references. Fortunately, it appeared that gelatin itself can provide a good reference system for unordered structure. We will discuss this in the following section.

New reference spectra for unordered structure

It has been argued that the spectrum such as observed for denatured gelatin is the characteristic CD-spectrum of proline-rich poly β-turn structures, which occur in proline-rich sequences [48–50]. In an attempt to clarify this we performed some measurements on proteins and peptides with primary structures that are very different from collagen and gelatin (see Fig. 4), for example the signal peptide of the bacterial outer membrane protein prePhoE [40, 51]. The CD-spectrum of this peptide is very similar to that of denatured gelatin, although the peptide contains no imino acid, and only a single glycyl residue. Interestingly, recent experiments with idealized signal peptides that are devoid of prolyl and glycyl residues also yielded spectra that were closely similar to the spectrum of denatured gelatin [41]. Clearly, prolyl residues are not essential for the induction of this type of spectra, and the idea that such spectra are exclusively associated with poly β-turn structures is probably not valid. This type of spectra is often obtained with polyamino acids under conditions that favour unordered structure: at elevated temperature, or in the presence of chaotropic ions [29–33]. Essentially the same CD spectra were obtained with apocytochrome c (the precursor of cytochrome c that lacks the heme

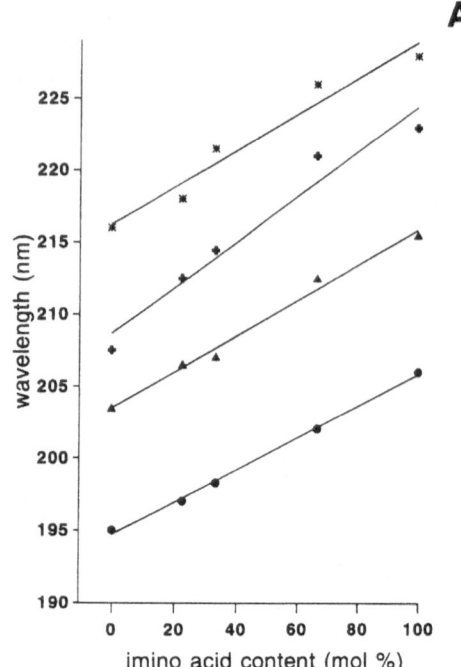

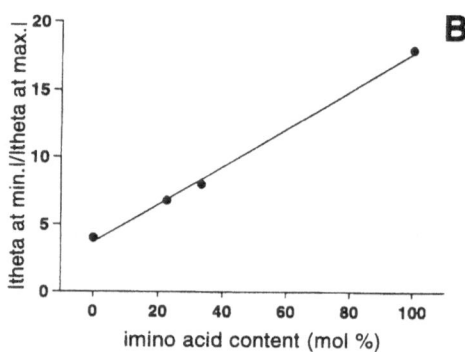

Fig. 3A Wavelength shift of the minimum (●), "midpoint" (▲), zero crossing (+) and maximum (∗) in the CD spectra of polypeptides with polyProline II-like secondary structure (i.e. with 3_1 helix conformation), as a function of the imino acid content (mol proline + hydorxyproline per mol amino acid in polypeptide). **B** Ratio of the absolute theta values at the minima and maxima defined in Fig. 3A, as a function of the imino acid content. The data were obtained from the following polypeptides: 0% imino acid residues: 0.01% (w/v) polyLysine at pH 5.7 and 22 °C; 22.5% imino acid residues: 0.0007% (w/v) collagen at pH 3.5 and 4 °C; 33% imino acid residues: 0.1–0.2% (w/v) poly(Gly-Pro-Ala) in water or buffer (pH 3.5) at 4 °C, or in ethylene glycol/water (2/1, v/v) at −112 °C (data in accordance with refs. [19, 37]), and in addition: 0.1% (w/v) of a random copolymer of 33 mol% proline, 33 mol% threonine, 33 mol% arginine in water at 5 °C; 66% imino acid residues: 0.05–0.2 % (w/v) poly(Prolyl-Glycyl-Proline) in water at 4 °C; 100% imino acid residues: 0.05% (w/v) polyProline in water at 4 °C

group) (Fig. 4), with prePhoE in urea, and with the bacterial protein SecA at elevated temperatures [52]. Also sodium caseinate, a mixture of non-globular proteins with only low contents of α-helix and β-structure [53–56],

Progr Colloid Polym Sci (1996) 102:9–14
© Steinkopff Verlag 1996

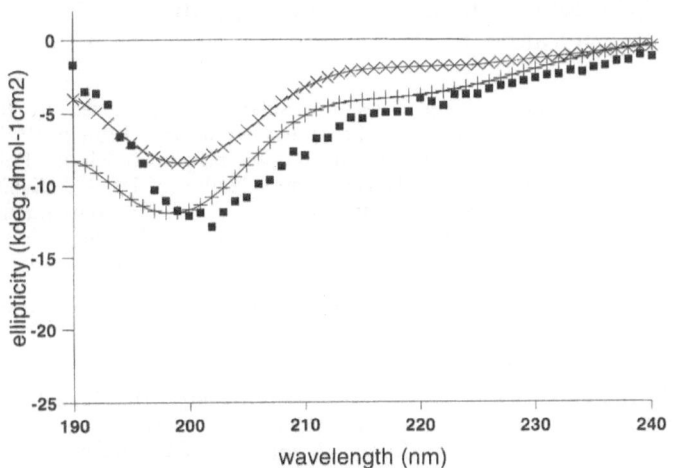

Fig. 4 CD spectra of polypeptides that are probably predominantly in a random conformation: prePhoE signal peptide (+), apocytochrome c (×) and sodium caseinate (■). The conditions were: signal peptide (0% imino acid residues): 0.00125% (w/v) in water at 22 °C; apocytochrome c (3.8 mol% proline residues): 0.05% (w/v) in 10 mM Tris-HCl buffer, pH 7.0 at 60 °C; sodium caseinate (overall average: 10.9 mol% proline residues): 0.05% (w/v) in water at 22 °C

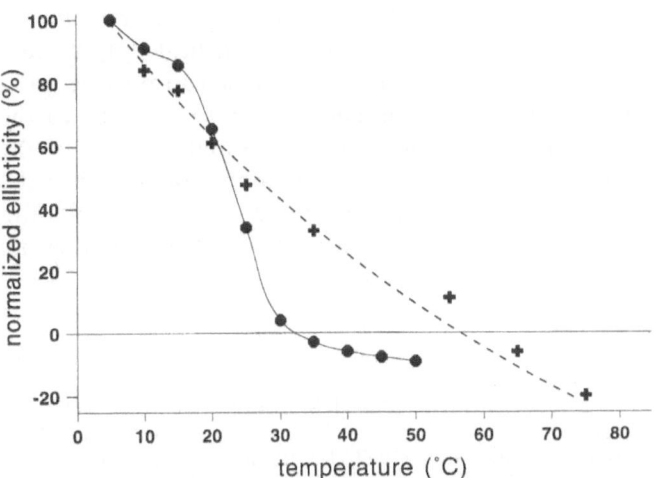

Fig. 5 Melting of the helices of gelatin (●) and polyLysine (+), monitored by the temperature-dependence of the ellipticity at 220 nm. For the sake of comparison, the ellipticities (see for example Figs. 1, 2) are normalized to the values at 4 °C. The conditions were: gelatin: 2.5% (w/v) in phosphate buffer, pH 7.0; polyLysne: 0.1% (w/v) in acetate buffer, pH 5.6. The temperature was increased stepwise from 4 °C to the indicated temperatures, and the measurements were performed approximately 5 min after completion of the temperature adjustment.

yielded a spectrum very similar to that of denatured gelatin and other putatively unordered polypeptides (Fig. 4). The overlap with the spectra of the signal peptide, apocytochrome c, or denatured gelatin was found to amount to 93–97%, while 2–5% was accounted for by α-helix, and 1–3% by β-strand. The most probable explanation for these findings is that CD spectra like those of denatured gelatin, and those shown in Fig. 4, are characteristic of unordered polypeptide structure.

Thus, well-denatured gelatin or collagen can be considered good representatives of unordered polypeptide structure. As such, they can be used as random coil references with the appropriate prolyl content for gelatin systems. This allows to discern triple helical and random structures in gelatin solutions and gels by circular dichroism, using native and denatured collagen as references. For example, we could calculate that the gelatin samples of Fig. 1 contained 5 and 79% helical structure (95 and 21% random coil), at 40° and 4 °C, respectively.

Multiple and single helices

It appeared to be impossible to discern triple, single and other helical structures by the form of the CD spectra, but the temperature dependence of the spectra may provide some insight into this matter. In the past, the melting behaviour of the helix-conformation was investigated using optical rotatory dispersion (see for example refs.

[10, 57]). More recently this type of experiment has been performed with CD on collagen-like peptides [58, 59]. The gradual decrease of the positive CD around 215 nm, observed for polyLysine (Fig. 5) and other polypeptides [29, 31–33], and indicating a gradual melting of the helix conformation, is characteristic for single helices. Figure 3 shows that gelatin at high concentration (2.5%) exhibits a cooperative melting behaviour, indicating the presence of many multiple-helical structures. Preliminary experiments with gelatin at very low concentrations (0.005%, w/v) point to a major contribution of single helices, rather than multiple helices (data will be published later). Currently we investigate the conformational consequences of the pH, ionic strength, etc. for different gelatin batches.

Conclusions

Our present work shows that CD spectroscopy is a valuable tool for the analysis of the molecular behaviour of gelatin systems. It can be applied to dilute and concentrated solutions (Figs. 1, 5), but also to gels and dry films (not shown). The commonly-used reference CD spectra for unordered polypeptide structure cannot be applied to gelatin and other protein or polypeptide systems with a high content of unordered structure. It appeared that for the case of gelatin, denatured collagen or gelatin can well be used as references for unordered structure (compare with Fig. 4). Native collagen or 3_1 helical polypeptides with a prolyl content of 20–30 mol%, can provide good reference spectra for the helical conformation. Since the prolyl content influences the shape and position of the spectra (Fig. 3), only reference polypeptides with the appropriate prolyl content should be used. Single and multiple helices can be discerned by their different thermal behaviour (i.e. by the degree of cooperativity of the helix-to-coil transition).

Acknowledgement This work was financially supported by the Dutch Innovation Research Program on Industrial Proteins (IOP-ie), project IIE92011. The gelatin was a kind gift of Gelatine Delft B.V., Delft, the Netherlands. The authors thank Prof. Dr. B. de Kruijff and Dr. J.A. Killian (University of Utrecht), Dr. Ir. J.B. Bouwstra and Ing. J. Olijve for fruitful discussions and valuable criticism.

References

1. Rich A, Crick FHC (1961) J Mol Biol 3:483
2. Ramachandran GN (1967) In: Ramachandran GN (ed) Treatise on Collagen, Vol 1. Academic Press, New York, pp 103–183
3. Bella J, Eaton M, Brodsky B, Berman HM (1994) Science 266:75
4. Piez KA, Carrillo AL (1964) Biochemistry 3:908
5. Harrington WF, Von Hippel PA (1961) Arch Biochem Biophys 92:100
6. Creighton (1993) Proteins – Structures and molecular properties. WH Freeman & Co, New York, pp 171–182
7. Grathwohl C, Wüthrich K (1976) Biopolymers 15:2025
8. Sarkar SK, Young PE, Sullivan CE, Torchia DA (1984) Proc Natl Acad Sci 81:4800
9. Harrington WF, Rao NV (1970) Biochemistry 19:3714
10. Harrington WF, Karr GM (1970) Biochemistry 9:3725
11. Hauschka PV, Harrington WF (1970) Biochemistry 9:3734
12. Hauschka PV, Harrington WF (1970) Biochemistry 9:3745
13. Hauschka PV, Harrington WF (1970) Biochemistry 9:3754
14. Te Nijenhuis K (1981) Colloid Polym Sci 259:107
15. Te Nijenhuis K (1990) In: Burchard W, Ross-Murphy SB (eds) Physical networks – Polymers and gels. Elsevier Applied Science, London, New York, pp 15–33
16. Te Nijenhuis K (1991) Macromol Chem, Macromol Symp 45:117
17. Djabourov M (1988) Contemp Physics 29:273
18. Ross-Murphy SB (1992) Polymer 33:2622
19. Heidemann E, Roth W (1982) Adv Polym Sci 43:145

20. Engel J (1962) Z Physiol Chem 328:94
21. Benguigui L, Busnel JP, Durand D (1991) Polymer 32:2680
22. Tiffany ML, Krimm S (1969) Biopolymers 8:347
23. Woody RW (1977) J Polym Sci Macromol Rev 12:181
24. Woody RW (1992) Adv Biophys Chem 2:37
25. Harrington WF, Rao NV (1967) In: Ramachandran GN (ed) Conformation of biopolymers. Academic Press, London, New York, pp 513–531
26. Wu CC, Komoroski RA, Mandelkern L (1975) Macromolecules 8:635
27. Rabanal F, Ludevid MD, Pons M, Giralt E (1993) Bioploymers 33:1019
28. Engel J, Kurtz J, Katchalski E, Berger A (1966) J Mol Biol 17:255
29. Tiffany ML, Krimm S (1972) Biopolymers 11:2309
30. Tiffany ML, Krimm S (1968) Biopolymers 6:561
31. Drake AF, Siligardi G, Gibbons WA (1988) Biophys Chem 31:143
32. Dukor RK, Keiderling TA (1991) Biopolymers 31:1747
33. Makarov AA, Adzubei IA, Protasevich II, Lobachov VM, Fasman GD (1994) Biopolymers 34:1123
34. Dölz R, Heidemann E (1986) Biopolymers 25:1069
35. Thakur S, Vadolas D, Germann HP, Heidemann E (1986) Biopolymers 25:1081
36. Heidemann E (1987) Das Leder 38:81
37. Brown FR, Carver JP, Blout ER (1969) J Mol Biol 39:307
38. Greenfield N, Fasman GD (1969) Biochemistry 8:4108
39. Reddy GL, Nagaraj R (1989) J Biol Chem 264:16591
40. Keller RCA, Killian JA, de Kruijff B (1992) Biochemistry 31:1672

41. Izard JW, Doughty MB, Kendall DA (1995) Biochemistry 34:9904
42. Killian JA, Keller RCA, Struyve M, de Kroon AIPM, Tommassen J, de Kruijff B (1990) Biochemistry 29:8131
43. Fisher WR, Taniuchi H, Anfinsen CB (1973) J Biol Chem 248:3188
44. Eastoe JB (1955) Biochem J 62:589
45. Smith CR (1921) J Am Chem Soc 43:1350
46. De Wolf FA, Krab K, Visschers RW, de Waard JH, Kraayenhof R (1988) Biochim Biophys Acta 936:487
47. Yang JT, Wu CSC, Martinez HM (1986) Meth Enzymol 130:209
48. Tatham AS, Drake AF, Shewry PR (1985) Biochem J 226:557
49. Urry DW (1987) J Prot Chem 7:1
50. Fontenot JD, Tjandra N, Ho C, Andrews PC, Montelaro, RC (1994) J Biomol Struct Dynam 11:821
51. Nouwen N, Tommassen J, de Kruijff B (1994) J Biol Chem 269:16029
52. Shinkai A, Mei LH, Tokuda H, Mizushima S (1991) J Biol Chem 266:5827
53. Creamer LK, Richardson T, Parry DAD (1981) Arch Biochem Biophys 211:689
54. Farrell HM Jr, Brown EM, Mumosinski TF (1993) Food Struct 12:235
55. Sawyer L, Holt C (1993) J Dairy Sci 76:3062
56. Holt C, Sawyer L (1993) J Chem Soc Faraday Trans 89:2683
57. Von Hippel PH, Wong KY (1963) Biochemistry 2:1399
58. Brodsky B, Li MH, Gwyne Long C, Apigo J, Baum J (1992) Biopolymers 32:447
59. Venugopal MG, Ramshaw JAM, Braswell E, Zhu D, Brodsky B (1994) Biochemistry 33:7948

Progr Colloid Polym Sci (1996) 102:15–18
© Steinkopff Verlag 1996

V. Sovilj
P. Dokić

Influence of shear rate on gelation of biopolymers solution

Dr. V. Sovilj (✉) · P. Dokić
University of Novi Sad
Faculty of Technology
Bul. Cara Lazara 1
21000 Novi Sad, Yugoslavia

Abstract The gelation process of mixed biopolymer solution under the shear rate was investigated. To this aim 3% solution of gelatin and sodium carboxymethylcellulose (NaCMC), with the gelatin/NaCMC ratio 4/1 was prepared. Rheological measurements were performed using rotational viscometer with coaxial cylinders ("Haake", Germany) at the temperature of 20 °C. Continuous loop hysteresis method was applied. Flow curves of the gels of different ages were obtained after shearing up to D_{max}, 10 min shearing at D_{max}, and down shearing. The gels were also subjected to steady shear at constant shear rate in order to get shear stress–time dependence.

The fresh gel shows rheopexy, i.e., antithixotropy, shear stress increases under steady shearing. During the gel aging structural links are establishing in time, and the gels show thixotropy and then rheopexy. When the gel is subjected to a continuous shearing two processes are taking place simultaneously, link disruption, leading to a shear stress decrease and link setting up, leading to a shear stress increase. The whole effect, depending on which one of the processes prevails during steady shearing, is either rheopexy or thixotropy. A balance between disruption and growth of gel structure is established at a particular gel age.

Key words Gelatin – sodium carboxymethylcellulose – gelation – thixotropy – rheopexy

Introduction

Rheological properties of biopolymer solution are influenced by a variety of factors, such as biopolymer concentration, molecular mass, electrical charge of ionic groups, solvent nature, temperature, pH value, presence of interacting materials, etc. Many biopolymers solutions can undergo gelation due to ability of crosslinking of biopolymer molecules by secondary bonds (hydrogen, hydrophobic, or electrostatic bonding), rather than by covalent bonds. This leads to creation of a network throughout the system. This establishing of bonds and weak structural links is a kinetic process dependent on thermal history leading to growth of three-dimensional network within the system, i.e., to formation of a gel. During this stage, a variety of structures can develop by varying the thermal history of the gel. Gelation process increases viscosity and changes rheological properties of the system [1–3]. Once formed, the gel undergoes structural changes due to further crosslinking of biopolymer flexible chains. The rate of the crosslinking is the highest just after the gel point is reached (fresh gels) [4].

The gelation can be accelerated or slowed down by various factors. In this research influence of shear rate on the gelation of mixed biopolymer gelatin and sodium carboxymethylcellulose (NaCMC) solution is examined. At certain conditions gelatin and NaCMC interact forming soluble complex which significantly increases viscosity of the system [5]. Furthermore, possibility of complex formation between two polyelectrolytes could affect network formation, i.e., gelation under the shear rate conditions. Investigation of gelation processes of gelatin/NaCMC complex under shear rate conditions, could be of importance for practical application of such systems in food and pharmaceutical industry.

Materials and methods

Gelatin (225 Bloom), basic processed, was obtained from "Sigma"-SAD, with isoelectric point 5.2 (viscometrically determined) as amphoteric polyelectrolyte, and sodium carboxymethylcellulose, 98% purity, degree of substitution 0.77, obtained from "Milan Blagojevic"-Lučani Yugoslavia as anionic polyelectrolyte, were used. Viscosity average molecular masses of gelatin and NaCMC were 181 000 g/mol and 147 000 g/mol respectively, determined using intrinsic viscosity methods and the Mark–Houwink relationship [5, 6].

In this investigation 3% (w/w) solution of gelatin and sodium carboxymethylcellulose (NaCMC), with the gelatin/NaCMC ratio 4/1 was prepared by heating the solution to 40 °C. The pH value of solution was 5.2, which is near the IEP of gelatin. Previous investigations showed that under these conditions the formed gelatin/NaCMC complex produces maximal increase in viscosity of the solution [7].

By cooling of the solution gelation was induced and when the temperature reached 20 °C (after 10 min of cooling in the measuring cylinder of the viscometer) the system was taken as a gel of the age of zero minutes. The other solutions after reaching 20 °C were kept at this temperature for 5, 10, 15, 20, 30 and 50 min respectively in order to obtain gels of different ages and after that, measurements were carried out.

Rheological measurements were performed using rotational viscometer RV20 ("Haake", Germany) with x–y–t plotter at the temperature of 20 °C. Continuous loop hysteresis method was applied [8]. Shear stress τ (Pa) was determined with continually changing shear rates D (s^{-1}) from zero ($D = 0$) to the maximal one (D_{max}) and then the reverse (after 10 min shearing at D_{max}), by an acceleration/deceleration of the viscometer rotor revolution of 400 min^{-2}. Obtained flow curves form a hysteresis loop.

Loop area was characterized by coefficient of thixotropy K, defined elsewhere [9, 10] as:

$$K = \frac{1}{n}\sum_{1}^{n} \Delta\tau_i \tag{1}$$

where n is the number of differences $\Delta\tau_i = \tau_i' - \tau_i''$ of "up" (τ') and "down" (τ'') flow curves for the given D_i values. Negative value for coefficient K characterizes rheopectic behavior of a system i.e., antithixotropy.

Up-flow curves of the gels of increasing ages are difficult to express by an appropriate equation, but down-flow curves which describe somewhat the equilibrium state can be presented by the known Ostwald–Reiner power law:

$$\tau = CD^n \tag{2}$$

where coefficient C is the measure of consistency of a system and n is degree of non-Newtonian behavior.

The gels were also subjected to steady shear at constant shear rate (D_{max}) in order to get shear stress–time dependence. For gels of different ages a set of shear stress–time curves was obtained.

Results and discussion

Flow curve of fresh gels show rheopexy, i.e., antithixotropy. Under steady shear at D_{max} the gels show increasing shear stress and the down-curve has higher shear stress values which correlated with the up-curve in the whole range of D_i. Such rheological behavior of fresh gelatin/NaCMC gels indicates that flexible chains uncoil and there is orientation of chains in shear [3, 11]. This leads to exposing new junction points and the possibility of their linking, which initiates growth of biopolymer network, i.e., gelation in shear and rheopexy. During gel ageing structural links are established in time, and gels show first thixotropy and then rheopexy (Fig. 1). In the aged gels during shearing weaker structural links were interrupted first and the network was partially disrupted whereas further shearing brought about the orientation of flexible chains and establishment of new links greater in number, as well as to rebuild the stronger network and viscosity increase, i.e., rheopexy.

The older the gel, the greater the number of the links building the network, and the transition from the thixotropy to the rheopexy is taking place at higher rate of shear (Fig. 1). Also, with increasing gel age effect of disruption of gel network becomes more pronounced and the thixotropic loop larger in area compared to the rheopectic one. Therefore at gel age of 10, 30 and 50 min coefficient K (Table 1) has positive value and increases with gel age because of prevailing thixotropic loop area. Also the parameters of down-flow curves C and n show that the consistency of the gels and the degree of non-Newtonian behavior increase with the gel age.

When the gels of different ages are subjected to a continuous shearing at constant shear rate, two processes are

Progr Colloid Polym Sci (1996) 102:15–18
© Steinkopff Verlag 1996

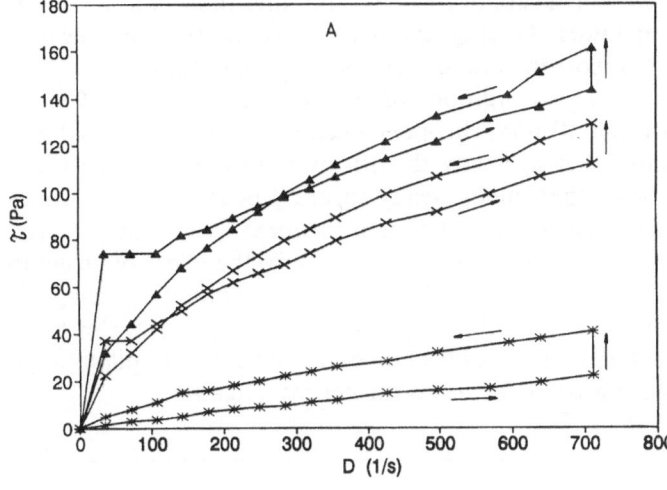

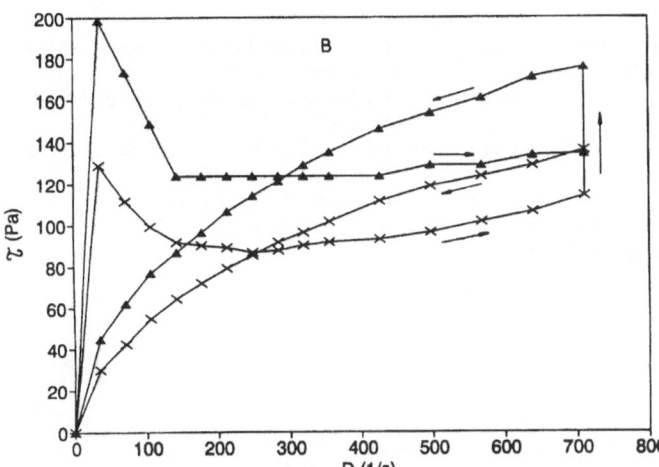

Fig. 1 Flow curve of 3% gelatin/NaCMC gel of different ages: A) fresh (*), 5 min (×) and 10 min (▲); B) 30 min (×) and 50 min (▲)

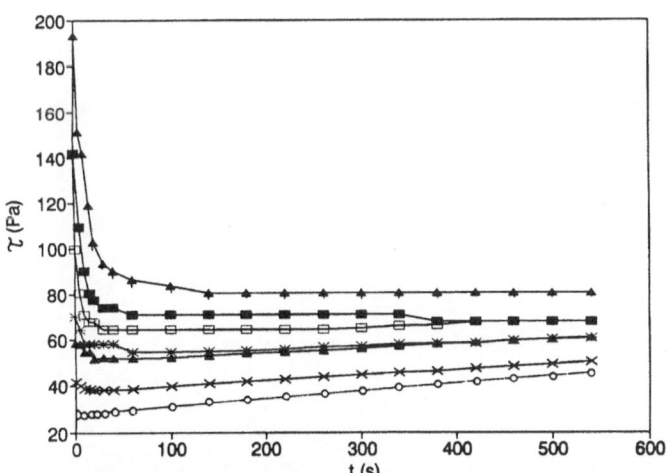

Fig. 2 Changes of shear stress with time of shearing under shear rate D = 936 s^{-1} for gelatin/NaCMC gel of different ages: fresh (○), 5 min (×), 10 min (▲), 15 min (*), 20 min (□), 30 min (■) and 50 min (▲)

Table 1 Coefficient K and parameters of the down-flow curves (Eq. (2)) for 3% gelatin/NaCMC gels of different ages

Gel age (minutes)	Parameter (Eq. (2))			Coefficient
	C	n	r	K
fresh	0.532	0.629	0.9997	− 12.45
5	2.68	0.595	0.9988	− 6.61
10	4.58	0.542	0.9996	3.14
30	6.08	0.477	0.9978	9.51
50	8.78	0.462	0.9989	17.28

r – coefficient of correlation.

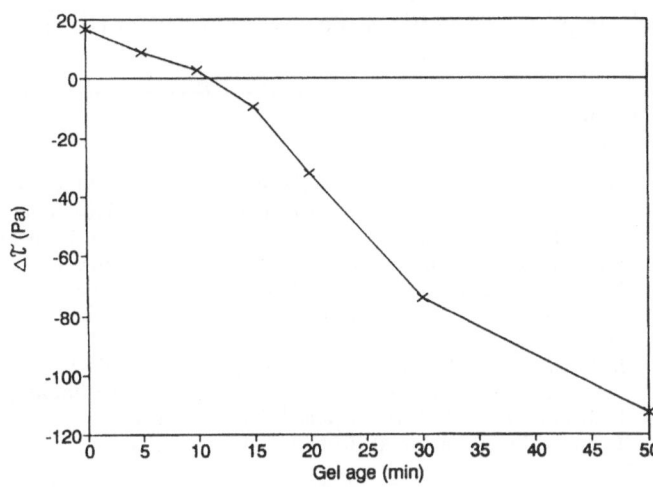

Fig. 3 The difference in initial and final value of shear stress ($\Delta\tau$) as a function of gel age, obtained at steady shearing

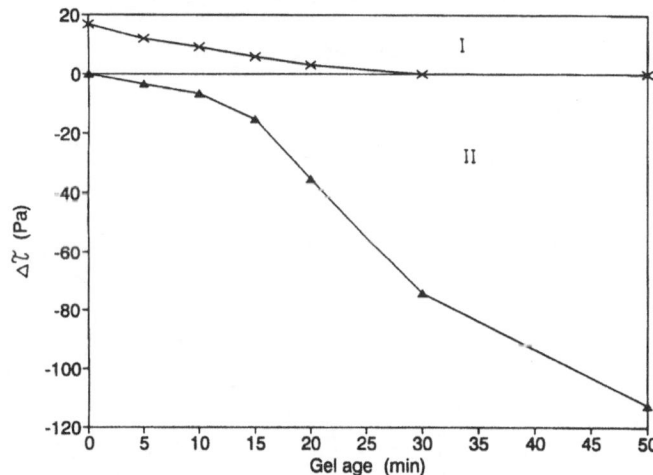

Fig. 4 The difference in shear stress $\Delta\tau$ at link setting up (I) and link disruption (II), as a function of gel age, obtained at steady shearing

taking place simultaneously, i.e., link disruption, leading to a shear stress decrease and link setting up, leading to a shear stress increase (Fig. 2), except fresh gel which show only link setting up.

The whole effect of shearing represented as the difference in initial (at time = 0) and final (at time = 540 s) value of shear stress ($\Delta\tau$), is either rheopexy or thixotropy, depending on which one of the processes prevails. The difference $\Delta\tau$ as quantitative measure of rheopexy is a positive one which decreases with the gel age and becomes negative when transition from rheopexy to thixotropy takes place (Fig. 3). A balance between disruption and growth of gel network during 10 min of shearing is established at gel age of about 12 min.

The differences $\Delta\tau$ caused by disruption and setting up of the gel network obtained from τ–t curves of different gel age are presented in Fig. 4. Positive values of $\Delta\tau$ (I) represent a measure of rheopexy and intermolecular crosslinking under shearing, and negative values (II) are a measure of thixotropy and disruption of gel structure.

Under conditions of steady shearing at $D = 936\,\mathrm{s}^{-1}$ during 10 min, rheopexy occurs until gel age of 20 min, after that age only thixotropy is observed. Since Fig. 1 shows that under steady shearing at $D = 712\,\mathrm{s}^{-1}$ during 10 min at gel age of 50 min rheopexy still occurs, whereas at higher shear rate ($936\,\mathrm{s}^{-1}$) only the effect of structure disruption occurs during the same time of shearing, the importance of shearing history and intensity of shear during measuring of flow curves are indicated. Explanation of such a behavior needs further investigations.

References

1. Ward AG, Courts A (1977) The Science and Technology of Gelatin. Academic Press, New York, London
2. Batdorf JB, Rossman JM (1973) In: Whistler RL (ed) Industrial Gums. Academic Press, New York, London, pp 695–729
3. Riihimaki TA, Middleman S (1974) Macromolecules 7:675–680
4. Ledward DA (1986) In: Mitchell JR, Ledward DA, Functional Properties of Food Macromolecules. Elsevier, London, New York, pp 171–201
5. Koh GL, Tucker IG (1988) J Pharm Pharmacol 40:233
6. Pouradier J, Vent AM (1952) J Chim Phys 49:391–398
7. Sovilj V, Dokić P, Trbović M (1995) 9th World Congress of Food Science and Technology, Budapest, 07.30–08.04, p 151
8. Dokić P, Djaković LJ, Sovilj V (1992) Kem Ind 31:447
9. Djaković LJ, Sovilj V, Milošević S (1990) Starch/Staerke 42:380–385
10. Djaković LJ, Milošević S, Sovilj V (1994) Starch/Staerke 46:266–272
11. Cascales JJL, Garcia de la Torre J (1990) Macromolecules 23:809–813

Progr Colloid Polym Sci (1996) 102:19–25
© Steinkopff Verlag 1996

G. Palazzo
M. Giustini
A. Mallardi
G. Colafemmina
M. Della Monica
A. Ceglie

Photochemical activity of the bacterial reaction center in polymer-like phospholipids reverse micelles

Dr. G. Palazzo (✉) · G. Colafemmina
M. Della Monica
Dipartimento di Chimica
Università di Bari
via Orabona 4
70126 Bari, Italy

M. Giustini · A. Mallardi
CNR
C. Studi Chimico-Flsici sull'Interazione
Luce-Materia
70126 Bari, Italy

A. Ceglie
Università del Molise
Facoltá di Agraria
86100 Campobasso, Italy

Absract An integral membrane protein, the photosynthetic bacterial reaction center (RC), has been incorporated in reverse micelle viscoelastic gels made of phosphatidylcholine and phosphatidylserine. Due to the dynamic nature of the gels, the use of a technique which shares the same timescale of the charge recombination is advised, in order to correlate the kinetic behaviour of the RC to the hosting-system properties. Self-diffusion and conductivity measurements have been used to investigate the properties of the model system lecithin/cyclohexane/water. The results indicate that such techniques can describe the properties of the system on a long characteristic time-scale. As a consequence, the kinetic behaviour of the RC has been studied by means of flash-spectrophotometry and related to the structural properties of the hosting gel, investigated by means of conductivity. The conductivity data are consistent with a water-induced sphere-to-rod transition of the phospholipid aggregates. Furthermore, increasing the ratio [water]/[lipid], a maximum in the hydrodynamic dimension of the giant worm-like reverse micelles is found. The experimental P^+ decay has been resolved into three exponential components which are strongly affected by the system composition. The functionality of the binding site Q_B is dependent on the ratio [water]/[lipid] supporting the hypothesis of a water role in the binding process.

Key words Charge-recombination – organogels – self-diffusion – membrane model

Introduction

Reverse micelles at moderately high values of surfactant concentration and mole ratio of water to surfactant (W_0) are generally believed to have a droplet-like structure.

The addition of small water amount to reverse micellar solutions usually induces a spherical growth of the aggregates, with no significant changes of the macroscopic solution properties, such as viscosity. However, Scartazzini and Luisi reported in 1988 a completely different behaviour for lecithin reverse micelles in a number of organic solvents [1]. These solutions can be transformed into transparent, highly viscous, thermodynamically stable, viscoelastic systems (organogels) by the addition of tiny amounts of water. In later years, widespread investigations, mainly due to Schurtenberger and coworkers, have firmly established the close analogy between the behaviour of these lecithin reverse micellar solutions and the classical polymer solution and have explained such polymer-like properties with a water-induced one-dimensional growth of the micellar aggregates into very long and flexible cylindrical reverse micelles [2–4].

Lecithin microemulsion gel has been used as model system which describes the dynamic properties of "equilibrium polymer" [3] and (when made of a biocompatible oil) as transdermal drug delivery system [5]. Furthermore, polymer-like phospholipid reverse micelles can be successfully used as membrane model, supplying a mimetic system of the hydrophobic core of the plasmatic membrane. From this point of view, the introduction of an integral membrane protein could be significant in order to investigate the relevance of the lipid–protein interaction on the protein properties. The high viscosity of the phospholipid based gels is a disadvantage in studying chemical reactions catalyzed by enzymes, the mixing of the reactants (both enzyme and substrate/s) being extremely difficult. Conversely, the organogels seem to be the ideal media for the study of photochemically-active proteins, such as bacterial photosynthetic reaction centers, due to their excellent optical transparency. The bacterial reaction center (RC) from the purple bacterium *Rhodobacter sphaeroides* is a transmembrane protein complex which contains three subunits (L, M, and H) and several cofactors, namely four bacteriochlorophylls (*bchl*), two bacteriopheophytins (I), two quinones (Q) and one non-haeme high-spin iron atom (Fe^{2+}) [6]. Two *bchl* form a dimer (P), which acts as the light-driven primary electron donor. The absorption of a photon promotes P to its excited state (P*). An electron is consequently transferred through I to the first quinone acceptor (Q_A which is located in a hydrophobic pocket of the protein) and subsequently to a second quinone molecule (Q_B) [7]. The radical anion Q_B^- is tightly bound to the protein; in contrast, Q_B is loosely bound so that the free exchange of quinone between the Q_B binding site of RC and the membrane pool is possible. In the absence of any exogenous electron donor to P$^+$, the charge recombination between P$^+$ and Q_B^- is observed. If, otherwise, the Q_B binding site is empty or in the presence of an inhibitor of the electron transfer between Q_A^- and Q_B, the light-induced charge separation and the successive recombination are limited to P$^+Q_A^-$. Since the negative charge is more stabilized when it resides on Q_B, the experimentally observed kinetics of P$^+$ decay depend on the parameters ruling the quinone exchange [8].

The polymer-like lecithin reverse micelles offer the opportunity to study the lipids mobility influence on the kinetic behaviour of the photoactive integral proteins. The successful attempt to solubilize the RC into lecithin microemulsion gels has been recently reported by our group [9, 10]. However, in order to preserve the photoactivity of the protein, the use of a mixture of phosphatidylserine (PS) and lecithin (PC) instead of pure PC has been necessary. In the present study, an attempt to correlate the properties of both the hosting gel and the guest protein with the system composition (lipid concentration and W_0) is presented.

Materials and methods

Soybean lecithin (epikuron 200) was a kind gift of Lucas-Meyer. Ubiquinone-10 (Q), phosphatidylethanolamine (PE), and phosphatidylserine (PS) (brain extract 85% of PS – sodium salt) were from Sigma. Cyclohexane and *n*-hexane were from Fluka. All the chemicals were of the highest purity available and were used without further purification.

RCs were isolated and purified from *Rhodobacter sphaeroides* R-26 strain as already described in the literature [11].

The phospholipids (PL) based organogels were prepared weighing the lipids in a screw cap glass vessel, dissolving them with the proper volume of organic solvents and then adding under stirring the water amount needed to obtain the desired W_0 [1].

The preparation of RC-containing phospholipid gels is described in detail elsewhere [9]; briefly, the RCs were first incorporated in lipid vesicles (PC/PE/PS 1:1:2) and then extracted in *n*-hexane in the presence of Mg^{++}. The water present in this RC/phospholipids organic solution was removed, incubating the extract for 6–8 h with anhydrous sodium sulphate (50 mg/mL), which was then removed by centrifugation.

In the experiments performed at increasing W_0, the desired amount of phospholipids (PC/PS 1:1 by weight) was first dissolved in a minimal amount of *n*-hexane and then added to the RC/phospholipids organic phase. Finally, the water was added to obtain the desired W_0's.

In the experiments performed at increasing lipid concentration and fixed W_0 ($W_0 = 4$) the lipids, as *n*-hexane solution, were gradually added to the RC/phospholipids organic phase together with the water needed to keep constant the W_0.

All the kinetic measurements have been performed in the presence of an ubiquinone excess (6 mM – in the experiments, at different lipid concentrations, this value decreases due to dilution).

Self-diffusion coefficients measurements have been carried out as described elsewhere [12]. The accuracy of the self-diffusion coefficients was within 2%.

Conductivity measurements were made by means of a CDM-83 conductimeter (Radiometer) at 73 Hz, using a thermostatted immersion cell (Amel 192 K1, cell constant 1.06 cm^{-1} at $T = 298 \pm 0.2$ K). The accuracy and reproducibility of the conductivity measurements were always better than 5%.

Progr Colloid Polym Sci (1996) 102:19–25
© Steinkopff Verlag 1996

In the case of cyclohexane gels, in order to perform the conductivity measurements, the addition of KCl (0.25 M) has been necessary since lecithin carries no net charge. A fixed amount of KCl solution was added (to obtain a W_0 of 2) and further additions of water were performed for those sample with higher values of W_0. It has to be noticed that the presence of the KCl reduces the W_0^{max} value to 8–10 at 298 K.

Conversely, in conductivity measurements on the PC/PS based organogels in n-hexane, no salt addition was needed since PS carries a negative net charge.

Flash-induced redox changes of the primary electron donor of the RC were monitored at 600 nm (a minimum in light-dark spectrum of the RC) with a single beam spectrophotometer of local design. Actinic flashes were provided by a xenon lamp (3.25 J discharge energy), screened through two layers of Wratten 88A gelatin filter, giving a light pulse of 4 μs duration at half-maximal intensity. Rapid digitization and averaging of the amplifier output was done by a LeCroy TR 8818 transient recorder equipped with a 128 K memory module (MM 8105) and controlled by an Olivetti M280 PC. Deconvolution of the charge recombination kinetics into multiple exponential decays was performed by computer routines based on the Marquardt algorithm.

Results and discussion

Although pure lecithin was found to be able to form gel in more than 50 different solvents [13], only two systems have been investigated in detail, namely PC/isooctane/water and PC/cyclohexane/water. In these systems the results coming from small-angle neutron scattering (SANS), static light scattering (SLS), and dynamic light scattering (DLS) experiments indicate a complex dependence of the micellar structure on both the lipid concentration and the W_0 [2]. At high lipid dilution a strong water-induced micellar growth, from relatively small spherical reverse micelles to giant worm-like aggregates, is observed. It was demonstrated by Schurtenberger and coworkers that these flexible giant reverse micelles are characterized by a contour length (L) which increases when W_0 is increased, keeping constant the values of both the cross-section diameter and the persistence length (l_p) of the aggregates. As soon as the concentration raises above a cross-over value c^*, these flexible, rod-like aggregates start to entangle, forming a transient network with static properties comparable to those of semidilute solutions of flexible polymers. A quantitative agreement between theoretical calculations performed on the basis of the renormalization group theory and the experimental data obtained for different static and dynamic properties is

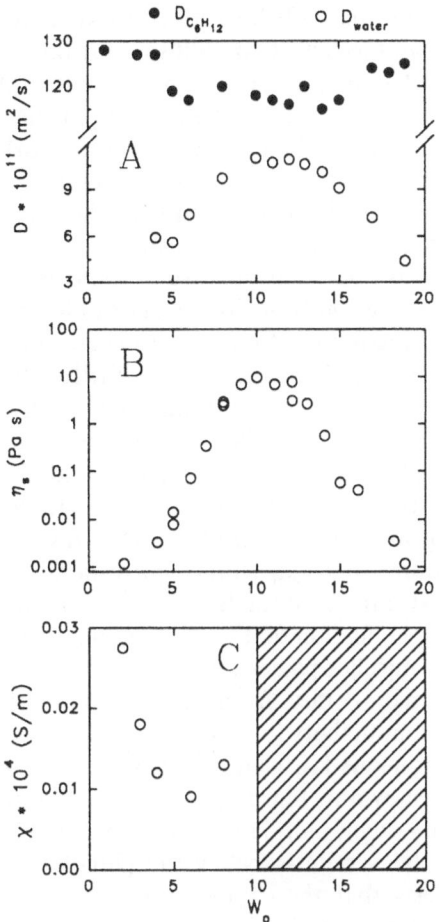

Fig. 1 Properties of the PC/cyclohexane gels. **A** Self-diffusion coefficients of water (hollow circle) and cyclohexane (filled circle) as function of W_0 ($C = 160$ mg/mL). **B** Zero Shear Viscosity as function of W_0 (data taken from ref. [2]). **C** Conductivity of KCl in a 160 mg/mL gels as function of W_0. The shaded area indicates the phase separation region

present in the literature [3]. On the other hand, as pointed out in ref. [3], this structural characterization of the system is restricted to static and/or dynamic properties observed on a very short characteristic timescale, where the finite lifetime of the aggregates can be neglected. A quite different behaviour has been reported using techniques characterized by longer timescales, such as the zero-shear viscosity (η_s). In this case, for both isooctane [14] and cyclohexane [2] based gels, η_s ranges over several order of magnitude and shows a characteristic bell-shaped curve. A similar trend has been observed by measuring other dynamic properties, by means of techniques characterized by a long characteristic timescale. In Fig. 1A the values of both the solvent and the water self-diffusion coefficients (measured by means of PFGSE-NMR experiments) for the system PC/cyclohexane/water at different W_0 are shown. Figure 1B shows the dependence of η_s on W_0 for the same system (data taken from ref. [2]). The PFGSE-NMR results suggest a different mechanism of diffusion for the oil and the water molecules, in agreement with the structural model of giant worm-like reverse micelles. The D_{oil} values are consistent with a solvent diffusion within a network characterized by an average mesh size which is function of W_0 [15]. The minimum in D_{oil} corresponds to the maximum L; for higher values of W_0 a decrease in the contour length is observed, as proposed in the case of rheological measurements [14]. The water behaviour is quite different. D_{water} values show a maximum of mobility in correspondence of the maximum in η_s. Furthermore, the NMR echo decays are strictly mono-exponential, ruling out the hypothesis of a restricted diffusion. The H_2O solubility in cyclohexane being negligible, the water diffusion should be due to the motion of the H_2O molecules along a tortuous curvilinear path inside the worm-like giant reverse micelles. The maximum in D_{water} should correspond to a maximum in the average contour length of the aggregates. Another strong indication that (on a long timescale) the dimension of the micelles passes through a maximum when W_0 is increased, comes from the conductivity measurements. The conductivity (χ) of cyclohexane based gels, prepared with aqueous KCl solution, has been measured. The χ values, shown in Fig. 1C, are extremely low and show a minimum when plotted as a function of W_0. This indicates that the system is below the percolation threshold and suggests that the charge carriers are the aggregates moving in a non-conductive oil. The conductivity behaviour of water in oil microemulsion at low water content has been explained on the basis of the charge fluctuation model [16, 17, 12], considering the energy needed to charge a body in a dielectric medium. A complete calculation, however, is present only in the case of a spherical geometry [16, 17] and cannot be used quantitatively in the case of polymer-like reverse micelles. Nevertheless, from a qualitative point of view, χ for tubular reverse micelles arranged in a random coil should be inversely proportional to the hydrodynamic radius of the aggregates. Consequently, the minimum in the plot of χ vs. W_0 should indicate a maximum in the dimension of the coil. It should be pointed out that this minimum matches neither with the maximum observed in the viscosity curve of pure lecithin gels (Fig. 1B), nor with the maximum observed in the D_{water} vs. W_0 plot (Fig. 1A). Moreover, the phase separation occurs at a lower W_0 than the unperturbed system (i.e.: without KCl) and the W_0 where the maximum viscosity is observed (as judged by eye) is around 5–6 instead of 10–12. These experimental evidences could be explained taking into account the effects exerted by KCl on the micellar properties. As found by Biocelli and coworkers on an analogous system (lecithin reverse micelles in benzene) by means of IR and ^{1}H-T_1

NMR measurements [18, 19], the presence of relatively high concentrations of KCl (0.1–1 M) induces a conformational modification of the phosphocoline moiety of the lecithin (from perpendicular to parallel to the long axis of the molecule) and an increase in the amount of water bound to the surfactant polar head. Since for lecithin organogels both these effects have been proved to be a crucial step in the gel formation [20, 21], the presence of the KCl would shift to lower W_0 all the gel features, phase separation and maximum viscosity included.

Since, as already told, the RC solubilization in the organogels requires the presence of the negatively charged phospholipid PS, a characterization of the hosting system PC/PS appeared necessary. It has to be pointed out that the PC/PS organogels show features typical of a viscoelastic solution, such as the response to an angular deformation (Weissenberg effect [22]). This evidence, together with the already reported ^{31}P-NMR spectra of the PC/PS organogels in n-hexane [10] allowed us to extend to this system the expression organogel. On the basis of this assumption, it was possible to determine the amount of water needed to obtain the maximum viscoealstic response as the W_0 value where maximum is the Weissenberg effect ($W_0 \approx 3$) and the W_0 value where phase separation occurs $W_0 \geq 7$). It should be remembered that in the RC the charge recombination occurs with a lifetime of the order of 1 s when the charge separation involves Q_B, and with a faster lifetime (0.1 s) in the presence of a $P^+Q_A^-$ recombination. Due to the dynamic nature of the gels, the use of a technique which shares the same timescale of the charge recombination is advised, in order to correlate the kinetic behaviour of the RC to the hosting-system properties. For this reason PC/PS in n-hexane gels have been investigated using conductivity. Since the PS is an anionic phospholipid, it is possible to measure the conductivity of the system over a wide range of surfactant concentrations (C) without adding any electrolytes. The results are summarized in Fig. 2. At high dilution and low W_0 ($W_0 = 0.5$ and 1), χ increases linearly with the PS concentration reflecting an increase in the aggregates concentration with constant dimensions. This is the behaviour expected for spherical or pseudo-spherical reverse micelles at constant W_0. However, for higher lipid concentration values a decrease in the slope of χ vs. C curve is observed. For samples at $W_0 = 1.8$ the two effects are comparable, resulting in a maximum in the χ vs. C plot at about 80 mg/ml. The increase of the W_0 to 2.8 levels-off the measured χ at high dilution, while for $C > 80$ mg/ml χ decreases when C increases (Fig. 2A). Also, in the case of PC/PS organogels a plot of χ vs. W_0 at fixed C shows a minimum in correspondence of the maximum of viscosity (Fig. 2B).

Although the conductivity behaviour of the n-hexane gels made of a mixture of PC and PS is not yet fully and

Fig. 2 Conductivity behaviour of PC/PS (1:1 in weight) n-hexane gels. **A** Conductivity values as function of total lipid concentration at different W_0. **B** Conductivity of a 160 mg/mL PC/PS gels as function of W_0

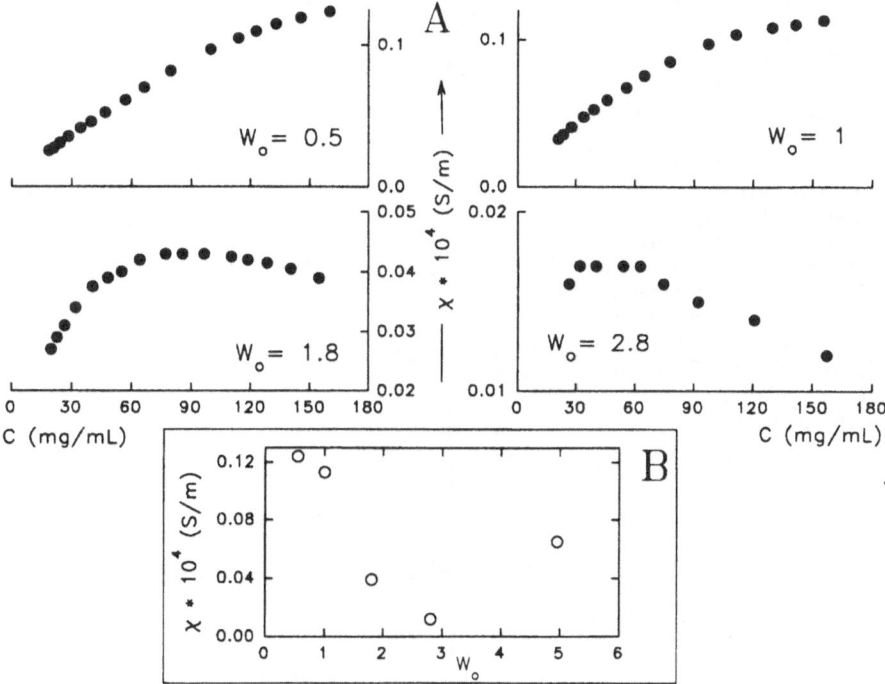

quantitatively understood, we believe that the above reported data suggests a water-induced sphere-to-rod transition of the micellar aggregates. Furthermore, it is evident that increasing C, at constant W_0, a decrease in the mobility of the aggregates, probably due to the formation of the transient network, is observed.

The kinetic behaviour of the RC hosted in the gel was investigated at different values of W_0 and C and reveals several interesting features. First of all, three exponentials are required to fit the decay of P^+: a fast phase with half-times $(t_{1/2})$ shorter than 100 ms, an intermediate phase characterized by $t_{1/2}$ values ranging from 150 ms to 700 ms, and a slow phase with $t_{1/2}$ longer than 1 s. The $t_{1/2}$ values obtained for the fast phase are in good agreement with those found in aqueous system [8]; furthermore, in the gels the addition of terbutryne, an inhibitor of the electron transfer from Q_A^- to Q_B, results in a monoexponential decay of P^+ with the same $t_{1/2}$ values (data not shown). On the basis of these considerations the fast phase can be attributed to the charge recombination occurring between P^+ and Q_A^- in RC lacking quinone at Q_B site. As a consequence, the intermediate and slow phases can be attributed to the $P^+Q_B^-$ recombination. A detailed discussion of the origin of the biexponential behaviour of the $P^+Q_B^-$-recombination of RC in organic environments is not the aim of this paper and can be found elsewhere [9, 10].

Both the relative amplitudes and the half-times of the three exponential phases are influenced by the W_0. In-

creasing the W_0 (at $C = 160$ mg/mL) the slowing down of the overall charge recombination rate is observed (see Fig. 3A). Figure 3B shows the relative amplitude of the fast, intermediate, and slow phases (A_f, A_i, and A_s respectively) as a function of W_0. It is clear that the increase of W_0 induces a linear decrease of A_f and a correspondent increase of A_s, suggesting an increase of the number of the Q_B sites accessible to the quinone dissolved in the organic bulk. Otherwise, the analysis of the half-times dependence on W_0 of the three exponential phases (as shown in Fig. 4) reveals a maximum in the $t_{1/2}$ in correspondence of $W_0 = 3$, indicating that the charge separated states are more stabilized under this condition. In particular, a comparison between the half-times of the fast phase at different W_0's reported in Fig. 4A and the conductivity data of Fig. 2B, suggests a correlation between the characteristic of the hosting system and the stabilization of the $P^+Q_A^-$ state. Recently, it has been reported that the rate constant for the charge recombination between P^+ and Q_A^- is increased by a factor of 2 or 3 in dehydrated phospholipid reverse micelles in n-hexane [23]. The same feature has been previously found in dried films of the native protein [24]. In order to test the existence of any correlation among the changes in both the stabilization of the charge separated states ($t_{1/2}$) and the functionality of the Q_B binding site (A_f) with the total amount of water and/or the W_0, a set of experiments at constant W_0 and different lipid concentration has been performed (Fig. 5).

Fig. 3 Kinetic behaviour of the bacterial reaction center in PC/PS n-hexane gels ($C = 160$ mg/mL). **A** Charge recombination kinetics of RC in gel at $W_0 = 4$(1) and $W_0 = 0$(2). The P^+ recovery has been fitted as a sum of three exponentials (solid lines). In the upper and lower part of the graph are shown the residuals. **B** Relative amplitudes of the exponential phases obtained by the deconvolution of the experimental traces, recorded in RC containing PC/PS n-hexane gels ($C = 160$ mg/mL)

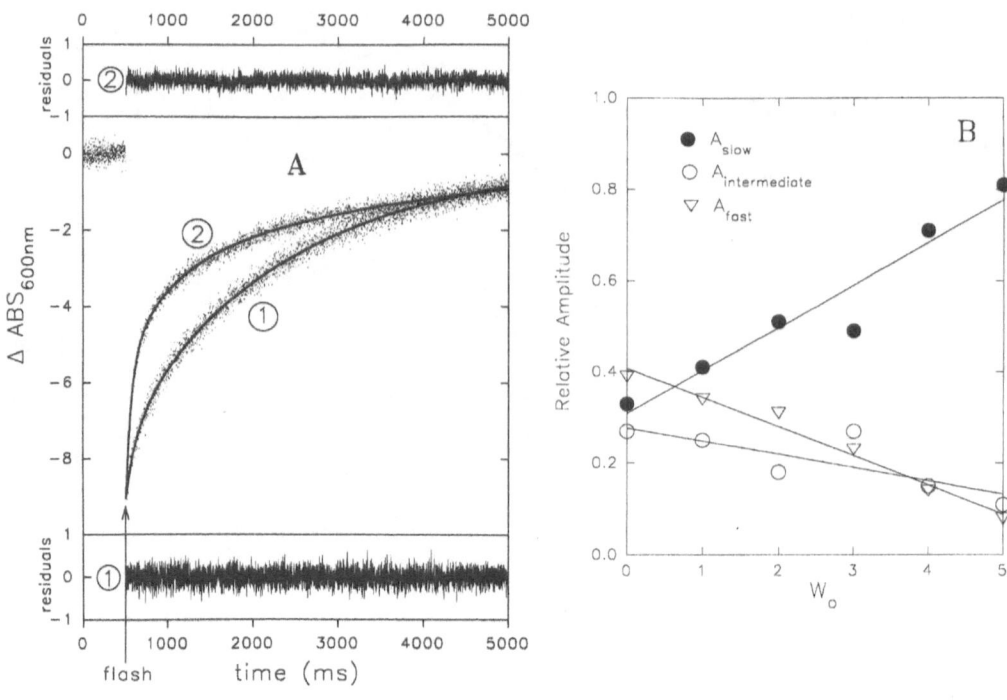

Fig. 4 W_0 dependence of the half-times of the fast (A), intermediate (B), and slow (C) phases of the P^+ recovery in RC containing PC/PS n-hexane gels ($C = 160$ mg/mL)

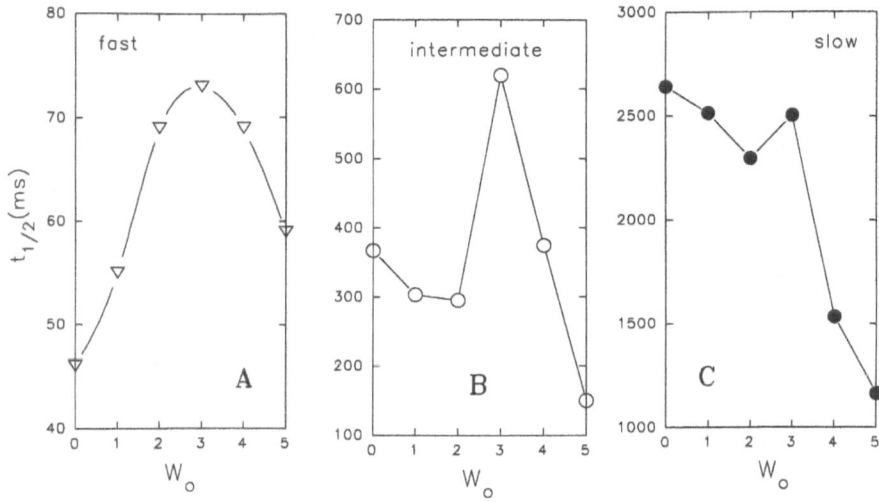

Since the kinetics in the organic solvents are strongly influenced by the quinone concentration [25] the analysis of the $t_{1/2}$ dependence on C is complicated by the changes occurring in this parameter (see Materials and Methods). Nevertheless, Fig. 5A clearly shows that at fixed W_0 the relative amplitude of the fast, intermediate, and slow phases are unaffected by the lipid and water concentrations. This information together with the data of Fig. 3B, indicates that the degree of organization of the water may play a role in the RC binding affinity for the unbiquinone molecules. This conclusion is in agree-

ment with the X-ray structure of the RC from *Rhodobacter sphaeroides* which reveals several water molecules buried in the core of the protein, some of them being well positioned to play a role in the binding process of the secondary quinone molecule, Q_B [6]. It should be mentioned that similar conclusions were recently proposed on the basis of investigations in aqueous systems at high osmolality [26].

A comparison of Fig. 4A and Fig. 5B indicates that the decrease in the stability of the $P^+Q_A^-$ state is related to the W_0 and not merely to the total amount of water. From the

Progr Colloid Polym Sci (1996) 102:19–25
© Steinkopff Verlag 1996

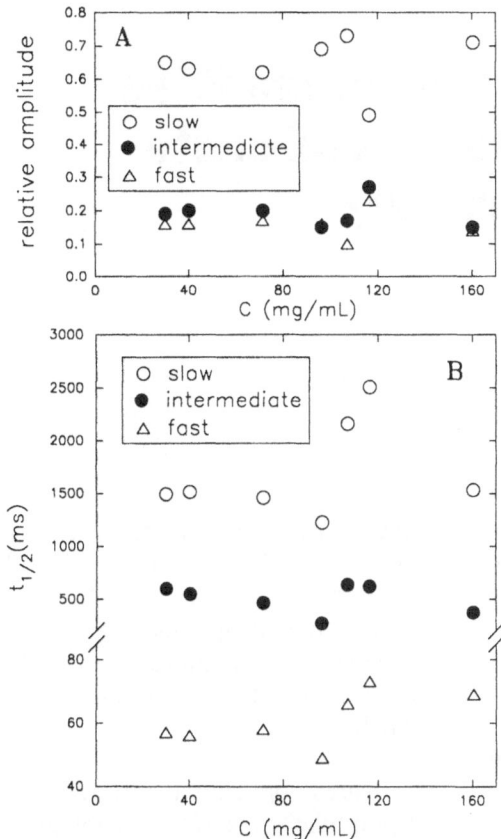

Fig. 5 Kinetic behaviour of RC in PC/PS n-hexane gels at fixed W_0 ($W_0 = 4$) and different total lipid concentration. **A** Relative amplitudes of the exponential components of P^+ reduction at different values of C. **B** Total concentration dependence of the half-times of the fast (hollow triangle), intermediate (filled circle) and slow (hollow circle) components of the P^+ recovery

x-ray structure of the RC there does not result any direct role for the water on the binding process of the ubiquinone at the Q_A site; this feature could be reasonably assigned to a change in the protein conformation.

As illustrated in Fig. 5B the $t_{1/2}$ of the three phases shows a monotonic dependence on C with an abrupt discontinuity at $C \approx 110$ mg/mL. This evidence suggests a structural transition of the hosting system, similar to that implied by the change in the slope of χ vs. C plot at lower W_0. At this stage of the work it is unfortunately impossible to prove this statement, lacking any information on the conductivity behaviour of the PC/PS gels at $W_0 = 4$.

Conclusions

Quite surprisingly, a close relation between the activity of a guest protein and the network properties of the host system has been found. Since some aspects of these interactions are not yet fully understood further measurements on both the system and the protein are required. The above reported data suggest that the polymer-like reverse micelles can be a powerful tool to investigate the lipid–protein interactions and to clarify the role played by the water and by lipid dynamics on membrane protein activity.

Acknowledgement We are grateful to Proff. M. Giomini and G. Venturoli for their stimulating discussions and to CONSORZIO INTERUNIVERSITARIO PER LO SVILUPPO DEI SISTEMI A GRANDE INTERFASE (CSGI-Firenze) for the finantial support.

References

1. Scartazzini R, Luisi PL (1988) J Phys Chem 92:829–833
2. Schurtenberger P (1994) Chimia 48: 72–78 and references therein
3. Schurtenberger P, Cavaco C (1994) J Phys Chem 98:5481–5486
4. Schurtenberger P, Cavaco C (1994) Langmuir 10:100–108
5. Hong-Li W, Luisi PL (1991) Biochem Biophys Res Comm 177:897–900
6. Ermler U, Fritzsch G, Buchanan W, Michel H (1994) Structure 2:925–936
7. Feher G, Allen JP, Okamura MY, Rees DC (1989) 339:111–116
8. Shinkarev VP, Wraight CA (1993) In: Deisenhofer J, Norris JR (eds) The Photosynthetic Reaction Center. Academic Press, New York, pp 193–255
9. Agostiano A, Catucci L, Della Monica M, Mallardi A, Palazzo G, Venturoli G (1995) Bioelectrochem Bioenerg 38:25–33

10. Agostiano A, Catucci L, Colafemmina G, Della Monica M, Palazzo G, Giustini M, Mallardi A (1995) Gazz Chim Ital 125:615–622
11. Gray KA, Farchaus JW, Wachtveitl J, Breton J, Oesterhelt D (1990) EMBO J 9:2061–2070
12. Giustini M, Palazzo G, Colafemmina G, Della Monica M, Giomini M, Ceglie A (1996) J Phys Chem 100:3190–3198
13. Luisi PL, Scartazzini R, Haering G, Schurtenberger P (1990) Colloid Polym Sci 268:356–374
14. Schurtenberger P, Scartazzini R, Luisi PL (1989) Rheologica Acta 28:372–381
15. Cukier RI (1984) Macromolecules 17: 252–255
16. Eicke HF, Borkovec M, Bas-Gupta B (1989) J Phys Chem 93:314–318
17. Callay N, Chittofrati A (1990) J Phys Chem 94:4755–4756
18. Boicelli CA, Giomini M, Giuliani AM (1981) Spectrochim Acta 37A:559–561

19. Boicelli CA, Conti F, Giomini M, Giuliani AM (1983) Gazz Chim Ital 113: 573–577
20. Capitani D, Segre AL, Sparapani R, Giustini M, Scartazzini R, Luisi PL (1991) Langmuir 7:250–253
21. Capitani D, Rossi E, Segre AL, Giustini M, Luisi PL (1993) Langmuir 9:685–689
22. Weissenberg K (1974) Nature 159: 310–313
23. Warncke K, Dutton PL (1993) Proc Natl Acad Sci USA 90:2920–2924
24. Clayton RK (1978) Biochim Biophys Acta 504:255–264
25. Mallardi A, Angelico R, Della Monica M, Giustini M, Palazzo G, Venturoli G (1995) In: Mathis P (ed) Photosynthesis: from light to biosphere. Kluwer AP, Amsterdam, Vol I: pp 843–846
26. Larson GW, Wraight CA (1995) Photosynth Res Supplement 1:65

Progr Colloid Polym Sci (1996) 102:26–31
© Steinkopff Verlag 1996

L. Picton
G. Muller

Rheological properties of modified cellulosic polymers in semi-dilute regime: Effect of salinity and temperature

L. Picton · Dr. G. Muller (✉)
URA 500
CNRS "Polymères, Biopolymères
et Membranes"
Université de Rouen
76821 Mont Saint Aignan Cedex
France

Abstract The rheological properties of low molecular weight (LW) and high molecular weight (HW) hydrophobically modified hydroxyethylcellulose (HMHEC) have been studied in semi-dilute regime of concentration ($C > C^*$) and compared with their unmodified parent polymers (HEC). Flow behavior and viscoelastic properties of modified polymers as a function of salt (NaCl, KSCN), temperature (20–60 °C) and shear rate (0.1–1000 s^{-1}) are explained by the existence of hydrophobic interactions. These latter are reinforced in the presence of water structure makers (NaCl).

Key words Associative polymer – amphiphilic – viscosity – rheology – viscoelastic

Introduction

Many applications where the control of rheology of aqueous media is of primary importance make use of high molecular weight flexible polymers (as partially polyacrylamide), or rigid biopolymers showing viscosifying or gelifying properties (such as xanthan gum). The viscosity enhancement is the consequence of chain extension, physical entanglement of solvated chains and/or stiffness of polymer backbone. However, such systems show limitations in terms of salt tolerance, mechanical and/or thermal stability.

In this regard, alternative approaches involve the ability of a new class of amphiphilic Associative Water Soluble Polymers (AWSP) to develop high viscosities due to the establishment of intermolecular interactions.

Such polymers contain a small amount of hydrophobic groups (i.e., long alkyl chains) attached to the hydrophilic polymer backbone. They can be prepared by copolymerization of water soluble and hydrophobic monomers [1–3] or by chemical modification of parent polymer [4–6]. In dilute solutions, the presence of hydrophobic groups affects the viscosity behavior mainly as consequence of intramolecular association (low [η], high k') [7–11].

In semi-dilute domain of concentration above a critical concentration for overlapping of chains, the low shear viscosity significantly increases because of intermolecular association.

If the number of hydrophobic side chains is at least equal to three per macromolecule, the association can be viewed as a pseudo cross-linking and the establishment of a three-dimensional network becomes possible, leading to an apparent increase of molecular weight. Therefore when compared to their precursors, such associative polymers show very peculiar rheological and viscoelastic properties [12, 13]. Such physical associations are of weak energy and can be disrupted under shear. At high shear rate the viscosity finally reaches the viscosity of the parent polymer. If no degradation occurred associations are rebuilt with decreasing shear.

Synthesis and solution properties of ionic and non ionic synthetic AWSP have been extensively reported in the literature [1–4]. By comparison, AWSP of natural

Progr Colloid Polym Sci (1996) 102:26–31
© Steinkopff Verlag 1996

origin have focused less attention and reported data mostly concerning the singular behavior of ASWP/surfactant systems [14–16].

New environmental considerations as well as natural abundance make polysaccharides, such as cellulosic water soluble derivatives, attractive precursors. Our interest has been focused on the solution properties of hydrophobically modified hydroxyethylcellulose (HMHEC) as compared to its parent HEC. It has been reported that adding alcohol to HMHEC leads to the disruption of hydrophobic association [7, 15]. This means that the three-dimensional hydrogen-bonded structure of water is of primary importance in the association process. As suggested by Franks [17], hydrophobic interactions occur with a further ordering of water. Therefore, we can expect that ions can play a role according to their effect on the water structure. For these reasons, we have compared the effect of water structure breakers (KSCN) and water structure makers (NaCl) on the rheological properties of HMHEC.

Experimental section

Polymer samples

HEC and HMHEC were kindly supplied by Aqualon company under the trade names Natrosol and Natrosol Plus, respectively. The synthesis of HMHEC has been reported by Landoll et al. [3] and consists of a chemical grafting of C_{16} alkyl epoxide on hydroxyl groups of the parent polymer in alkaline isopropanol/water slurry at 80 °C. Little if any information is known about the distribution of hydrophobic side chains. Nevertheless, some papers report a random distribution [15]. We found that no degradation results from the modification process [11]. Two samples of HEC/HMHEC differ by their molecular weight (Mw) of about 480 000 g/mol for the high Mw sample (HW) against 46 000 g/mol for the lower Mw (LW). Both parent polymer HEC ("precursors") have a degree of molar substitution MS = 3.6 and were modified by chemically grafting about 0.6% w/w of C_{16} alkyl groups. This means that the number of bonded hydrophobic groups per molecule is about 10 for HW HMHEC and less than 2 for LW HMHEC.

Polymer solutions

All solutions were prepared at ambient temperature by dispersion of powdered dried sample and stirred 1 day. They were stored at 4 °C. Dilute and concentrated solutions were stirred respectively with magnetic and mechanical devices. Pure Milli-Q water was systematically used

for preparing the solutions and a bactericide (NaN_3) was added to prevent any degradation of the polymer. In dilute regime clarified solutions were used: they were obtained by successive filtrations through 8; 3; 1.2; 0.65 and 0.45 μm Millipore filters. After filtration, no loss of polymer was detected for the precursors and HMHEC LW, whereas about 50% were retained in the case of HMHEC HW. This indicates that particles of large size (aggregates or microgels) were present in HW HMHEC. Moreover, LALLS measurements indicated the presence of some aggregates in filtered LW HMHEC.

NaCl, KSCN and urea were obtained from Prolabo (France) and were used as supplied.

Rheology

Viscometric measurements were carried out using a Low Shear 30 (Contraves) Couette type viscometer. The shear rate range was between 10^{-2} and 10^2 s^{-1} and the temperature was controlled by a thermal bath circulation (Polystat from Bioblock Scientific).

Flow curves and oscillatory shear responses were determined using a Carri Med CSL 100 (Rheo) controlled stress rheometer fitted with a Peltier temperature control $+/- 0.1$ °C. All the measurements have been performed with a solvent trap to prevent any evaporation of solvent.

The storage modulus (G') and the loss modulus (G'') were determined over the frequency range 0.01–10 Hz at constant low applied stress using the 4 cm/2° cone/plate measuring geometry. The linearity of viscoelastic properties has been verified for all the investigated solutions.

Results and discussion

Plots of apparent viscosity (25 °C) as a function of the applied shear rate ($\dot{\gamma}$) for modified and parent HW solutions in 0.1 M NaCl are shown in Fig. 1. The polymer concentration (10 g/L) was above the critical overlapping concentration C^* of both HEC ($C^* \sim 3$ g/L) and HMHEC ($C^* \sim 1$ g/L). In these experiments the rate of shear was increased until the maximum of approximately 1000 s^{-1} was reached and then immediately decreased down to 0.1 s^{-1}. As clearly evidenced the low shear viscosity of HMHEC is much larger than that of its precursor. Moreover, the modified polymer exhibits a strong pseudoplastic behavior. Finally, whereas HEC instantly reverts to its initial at-rest viscosity after shear removal, the modified HMHEC requires time for return to its at-rest initial state. This explains the hysteresis loop typical for a thixotropic solution.

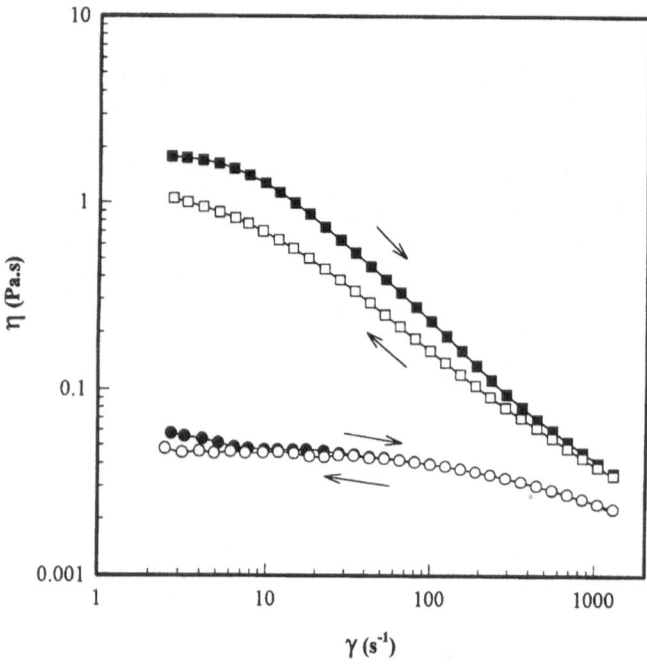

Fig. 1 Flow curves of HW HEC (circles) and HW HMHEC (squares): Cp = 1%, 0.1 M NaCl, 25 °C. Up to γ_{max}: filled symbols; down to γ_{min}: open symbols

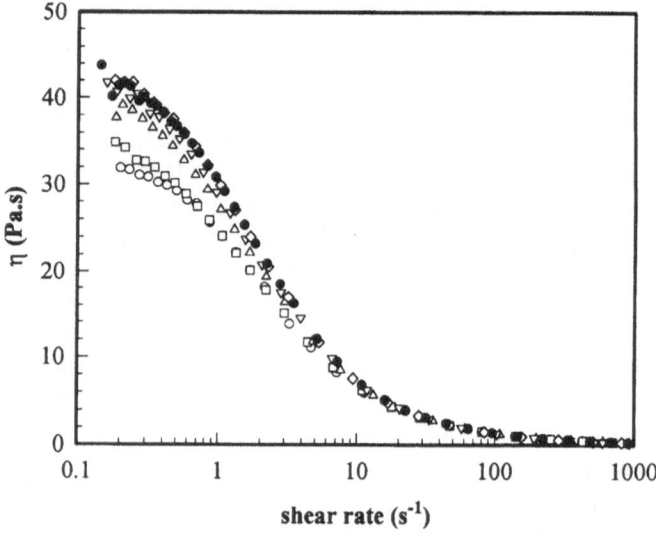

Fig. 2 Flow curves ($T = 20$ °C) of HW HMHEC (2%, 1 M NaCl) as a function of time at-rest (t) after shear disruption ($\gamma = 1000$ s^{-1}) of the network. ● initial flow; ○ $t = 0$; □ $t = 1'$; △ $t = 5'$ ▽ $t = 15'$; ◇ $t = 1$ h

Undoubtedly, intermolecular interactions can be held responsible for the rheological behaviour of HMHEC. These physical interactions are of weak energy and are easily disrupted as the shear rate increases. The hysteresis loop indicates that the three-dimensional structure does not build up again during the time-scale of these measurements.

The data reported in Fig. 2 show that the initial zero shear viscosity is totally recovered after 15 min at-rest after the shear removal. This means that the process is reversible and that after 15 min all the intermolecular associations are restored.

Salt influence

In Fig. 3 are reported the specific viscosities $(\eta - \eta_0)/\eta_0$ (measured at $\gamma = 10$ s^{-1}) of HMHEC LW and its parent HEC as a function of polymer concentration for different aqueous solution conditions. Pure water, lyotropic salt (NaCl), chaotropic salt (KSCN) and a well known chaotropic agent (urea) were used. By comparison with the precursor, the modified polymer shows strong upward curvature of the specific viscosity plots which reflects molecular association or aggregation. Whereas salt does not affect the rheological properties of the parent polymer [11], it is interesting to note that the curvature is much more pronounced in the sodium salt (a water structure

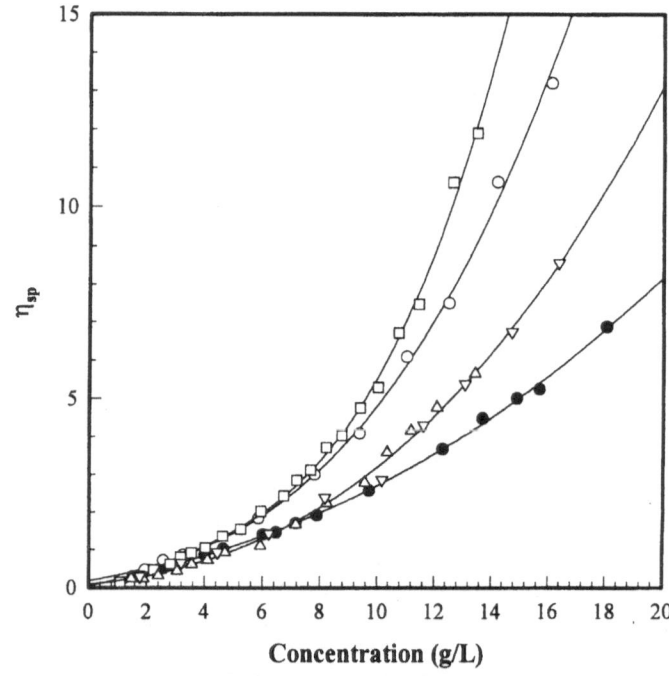

Fig. 3 Plots of specific viscosities as a function of polymer concentration: LW HEC (filled symbols), LW HMHEC (open symbols) for different solvent conditions: ○ water; □ 1 M NaCl; ▽ 1 M KSCN; △ 3 M urea

maker) than in KSCN or urea (water structure breakers). This indicates that the observed viscosity phenomena are the consequence of association among hydrophobic moities.

Frequency sweep experiments have been made to obtain informations upon viscoelastic properties of 2% HW

HMHEC. The storage (G') and loss (G'') moduli were measured at 25 °C as a function of frequency (0.01–10 Hz) under low applied stress (5 N/m²). The oscillatory results are given in Fig. 4 and useful information can be obtained from the critical frequency of cross over ω_c (where G' and G'' are equal). A critical time (τ_c) is associated with ω_c ($\tau_c = 1/\omega_c$) which corresponds to a "lifetime" of the junction of the network at a fixed stress [18]. It is clear that τ_c is longer in NaCl (5.5 s) than in KSCN (0.7 s), therefore confirming that gel properties are strengthened in presence of water structure makers that enhance hydrophobic interactions.

It is known that water has a three-dimensional hydrogen-bonded structure. When two hydrophobic molecules come into contact, the ordered water molecules around them are displaced. Hydrophobic interaction occurs with a further ordering of water.

Fig. 4 Storage (G' ●) and loss (G'' □) moduli versus frequency of HW HMHEC (2%, 25 °C, 5 N/m² of applied stress) in water (A), 1 M NaCl (B) and 1 M KSCN (C)

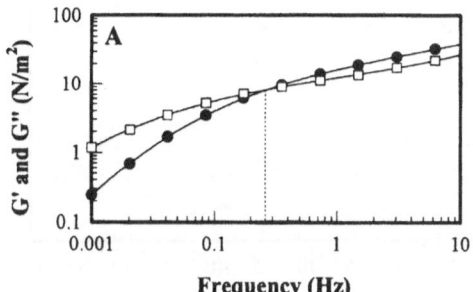

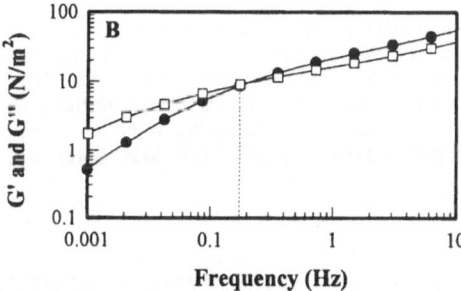

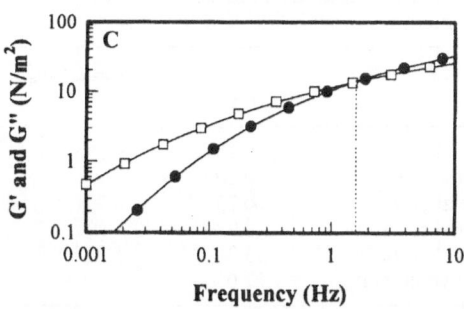

Effect of temperature

The temperature dependence of intrinsic viscosity ($[\eta]$) and Huggins constant (k') for LW HEC and HMHEC in pure water are reported in Table 1. For both samples the intrinsic viscosities decrease with increasing temperature. The systems differ by the change of k' versus temperature: whereas k' of precursor is slightly affected, a large increase is observed for the modified polymer (Fig. 5). As k' is related to polymer/polymer interactions, this indicates that increasing temperature strengthens hydrophobic interactions.

The same procedure as shown in Fig. 2 has been used (Fig. 6) to evaluate the time at-rest needed for the reorganization of the system HW HMHEC 20 g/L, 1 M NaCl at 60 °C after its disruption at high shear rate (1000 s⁻¹). A ten-fold decrease in the initial low shear viscosity is observed when the temperature rises from 20° to 60 °C (compare Fig. 2 and 6). HW HMHEC can be viewed as an entangled network of water soluble polymer chains in which hydrophobic groups are physically associated.

Table 1 Intrinsic viscosity ($[\eta]$) and Huggins constant (k') for LW HEC and HMHEC in pure water as a function of temperature

	HEC LW		HMHEC LW	
	$[\eta]$ mL/g	k'	$[\eta]$ mL/g	k'
25 °C	128	0.31	110	1.45
40 °C	100	0.75	91	2.22
60 °C	86	0.85	69	3.75

Fig. 5 Temperature dependence of the Huggins constant for LW HEC (○) and LW HMHEC (■) in pure water

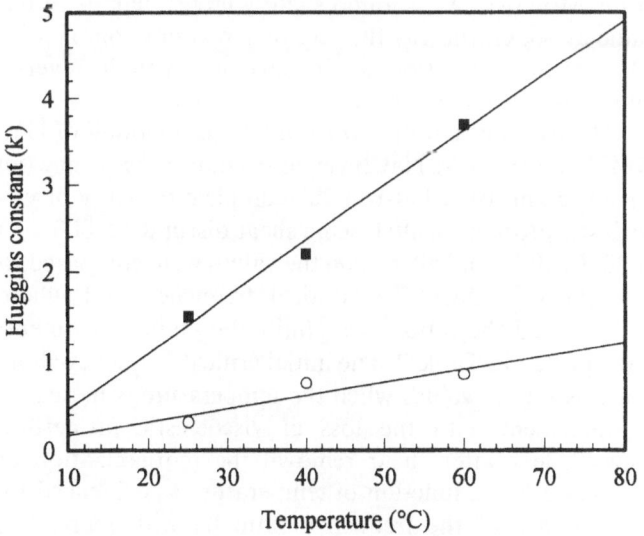

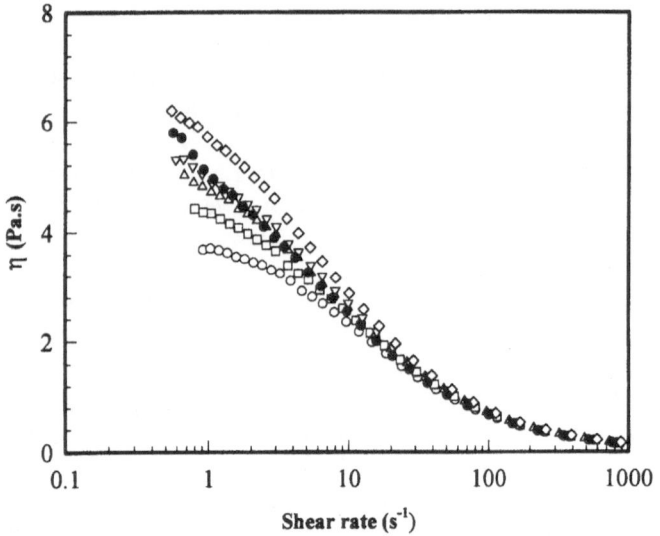

Fig. 6 Flow curves ($T = 60\,°C$) of HW HMHEC (2%, 1 M NaCl) as a function of time at-rest (t) after shear disruption ($\gamma = 1000\ \text{s}^{-1}$) of the network. ● initial flow; ○ $t = 0$; □ $t = 1'$; △ $t = 5'$; ▽ $t = 15'$; ◇ $t = 1$ h

It appears that the reformation of hydrophobic association is faster at higher temperature. Moreover, at higher temperature ($60\,°C$) and 1 hour at-rest after cessation of the shear treatment, the low shear recovered viscosity is higher than the initial one. Therefore it seems that increasing temperature makes a more organized network.

Oscillatory measurements have been made at $T = 20°$, $40°$ and $60\,°C$ to follow the reformation of the systems ($C = 2\%$, 1 M NaCl) HW HEC (Fig. 7a) and HMHEC (Fig. 7b) after being shear disrupted ($1000\ \text{s}^{-1}$). For both systems the elastic modulus (G') decreases with temperature. Instantaneous recovery is observed for the HEC precursor. On the other hand, the recovery is time dependent for the modified polymer and faster at higher temperature. Moreover, G' is found slightly larger than its initial value at $60\,°C$ whereas the recovery was not complete at $20\,°C$. Such observations are in agreement with the reversibility data above reported.

The dynamic storage (G') and loss (G'') moduli of HW HMHEC (2%, 1 M NaCl) versus excitation frequency (ω) have been measured at-rest 30' (complete recovery of viscoelastic properties) after being shear disrupted ($1000\ \text{s}^{-1}$) at $20\,°C$, $40\,°C$ and $60\,°C$, and the values were compared to the starting values. The critical frequency (ω_c) where $G' = G''$ and the ratio $r = \omega_c$ (initial)/ω_c (after treatment) are reported in Table 2. The initial critical frequency shifts towards higher values when the temperature is increased in agreement with the loss of viscoelastic properties. Nevertheless, after shear removal the reorganization of the network is a function of temperature as evidenced by the increase of the frequency ratio (r) with increasing

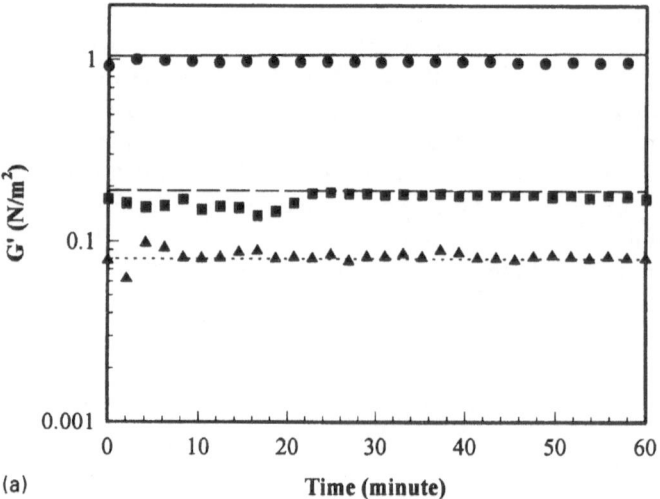

(a)

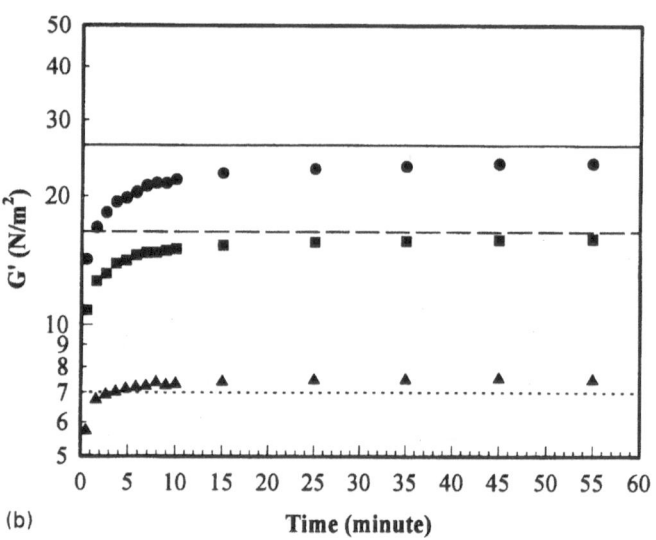

(b)

Fig. 7 Time dependence of the storage modulus (G') of a 2% polymer solution in 1 M NaCl as a function of temperature after a high shear disruption (15 min at $\gamma = 1000\ \text{s}^{-1}$). The applied stress was $3\ \text{N/m}^2$ and the measurement frequency was 1 Hz. $20\,°C$: —— initial state; ● recovery, $40\,°C$:——— initial state; ■ recovery, $60\,°C$: – – – – initial state; ◆ recovery, HW HEC: (a) and HW HMHEC (b)

Table 2 Critical frequencies (ω_c) of HW HMHEC 2% in 0.1 M NaCl as a function of temperature measured 30 min at-rest after a $1000\ \text{s}^{-1}$ disruptive shear treatment

		ω_c (Hz)	$r = \omega_c^1/\omega_c^2$
$20\,°C$	1: initial state	0.137	0.685
	2: recovered state	0.200	
$40\,°C$	1: initial state	0.73	0.79
	2: recovered state	0.92	
$60\,°C$	1: initial state	3.2	1.23
	2: recovered state	2.6	

Progr Colloid Polym Sci (1996) 102:26–31
© Steinkopff Verlag 1996

31

temperature. At high temperature (60 °C) the recovered ω_c is higher than initial ω_c, therefore in favour of a more structured recovered network.

Conclusion

Intermolecular hydrophobic interactions give rise to enhanced viscosity and viscoelastic properties in aqueous solution. Such associations are of weak energy and are easily disrupted under shear leading to a typical pseudo-plastic behavior. The time dependence of the reassociation process is evidenced by thixotropic properties.

As the three-dimensional hydrogen-bonded water structure induces hydrophobic interactions, we have observed that water structure makers (NaCl) reinforce such associations whereas water structure breakers (KSCN or urea) appear as limited factors of association.

High temperature (60 °C) leads to faster and much better reorganization of the associative network (after a total disruptive shear rate). These results are in agreement with the observations reported by Mc Cormick et al. on associative acrylamide/N-alkylacrylamide copolymers [3].

Acknowledgment The authors want to thank the Aqualon Company for the supplied samples of HEC and HMHEC and the CPR CNRS/ DIMAT for supporting this work.

References

1. Biggs S, Selb J, Candau F (1992) Langmuir 8:838–847
2. Schulz DN, Kaladas JJ, Maurer JJ, Bock J, Pace SJ, Schulz WW (1987) Polymer 28:2110–2115
3. Mc Cormick CL, Nonaka T, Johnson CB (1988) Polymer 29:731–739
4. Wang KT, Iliopoulos I, Audebert R (1988) Polym Bul 20:577–582
5. Landoll LM (1982) J of Polym Sci, Polym Chem Ed 20:443–455
6. Akiyoshi K, Degushi S, Morigushi N, Yamagushi S, Sunamoto J (1993) Macromol 26:3062–3068
7. Gelman RA, Barth HG (1986) In: Glass JE Ed., "Water Soluble Polymers: beauty with performance", Adv in Chem Series 213 pp 101–110, Am Chem Soc, Washington DC
8. Tanaka R, Meadows J, Phillips GO, William PA (1990) Carbohydrate Polymers 12:443–459
9. Magny B, Iliopoulos I, Audebert R (1991) Polymer Communications 32: 456–458
10. Bock J, Siano DB, Valint PL, Pace SJ (1987) Polym Mat Sci Eng 57:487–491
11. Picton L, Merle L, Muller G (1996) J of Polym Charact and Analysis 2:103–113
12. Sau AC (1987) Polym Mat Sci Eng 57:497–501
13. Goodwin JW, Hugues RW, Lam CK, Miles JA, Warren BCH (1989) In: Glass JE Ed., "Polymer in Aqueous Media: performance through association" Adv in Chem, Series 223 pp 365–378, Am Chem Soc, Washington DC
14. Tanaka R, Meadows J, Williams PA, Phillips GO (1992) Macromol 25:1304–1310
15. Varelas CG, Steiner CA (1990) In: "Absorbent Polymer Technology", pp 259–273, Elsevier Ed., Amsterdam
16. Sivadasan K, Somasundaran P (1990) Colloids and Surfaces 49:229–239
17. Franks F (1975) In: Franks F Ed., "Water: a comprehensive treatise" Vol 4, p 1, Plenum: New York
18. Leung PS, Goddard ED (1991) Langmuir 7:608–609

Progr Colloid Polym Sci (1996) 102:32–37
© Steinkopff Verlag 1996

M. Grisel
G. Muller

Rheological properties of schizophyllan in presence of borate ions

M. Grisel · Dr. G. Muller (✉)
URA 500
CNRS – Université de Rouen
76821 Mont-Saint-Aignan, France

Abstract The viscoelastic properties of schizophyllan, a neutral fungal polysaccharide, were studied in presence of small amounts of borate ions. We observed the ability to get physical gels as the result of the complexation between the diol sites of the biopolymer and the borate ions, leading to the establishment of a tridimensional network. Kinetic of gelation as well as the influence of physicochemical parameters were studied by way of rheological measurements.

Key words Schizophyllan – polysaccharide – borate ions – rheology – sol-to-gel transition – physical gel

Introduction

Aqueous solutions of poly(hydroxyl) compounds bearing *cis*-hydroxyl groups show a remarkable increase in viscosity or form gels when treated with crosslinking agents such as borate and titanium ions.

In this regard polyvinyl alcohol [1, 2, 3] or galactommanans [4, 5] e.g., guar gum/sodium borate complexation has been largely described together with the rheology of such systems.

Much evidence shows that gelation results from the attachment of boron to adjacent oxygens thus forming mono and/or dicomplexes. A network of physically linked macromolecules is formed as far as the concentration of crosslinked chains reaches a critical value.

As first pointed out by Boeseken and van Giffen in 1920 [6], and by Hermans in 1925 [7], alkalinity is necessary for effective complexation as far as the carbohydrate contains hydroxyl groups in a favourable position.

In the present work, we describe the rheological properties of aqueous solutions of schizophyllan in the presence of borate ions. The effect of both polymer and crosslinker concentration, pH and salinity influence were investigated. Kinetic data concerning schizophyllan – borate gels are also given.

Polymer presentation

Schizophyllan is a neutral extracellular polysaccharide produced by the fungus *Schizophyllan commune* [8]. Its chemical structure consists of linearly linked β-$(1 \rightarrow 3)$-D-glucose residues with one β-$(1 \rightarrow 6)$-D-glucose side chain for every three main chain residues, this structure being exactly the same as that of Scleroglucan [9] as indicated by NMR measurements (see Fig. 1a).

As shown by many workers, Schizophyllan chains in aqueous media adopt a triple-stranded helix conformation. This rodlike structure is responsible for its well-known ability to induce large viscosity enhancement of aqueous formulations, thus making the polymer well adapted for use in oil fields. A schematic representation of the triple helix has been proposed [10], consisting of a central cylinder having lateral side-chain D-glucose residues attached as indicated by Fig. 1b.

Furthermore, chemotherapeutic applications of this biopolymer as an antitumorous or antiinfectious agent [11, 12] opened a new field of interest to this molecule. Data have been reported showing Schizophyllan as an interesting molecule for drug delivery [13] and as a vehicle for ocular topical administration [14].

Progr Colloid Polym Sci (1996) 102:32–37
© Steinkopff Verlag 1996

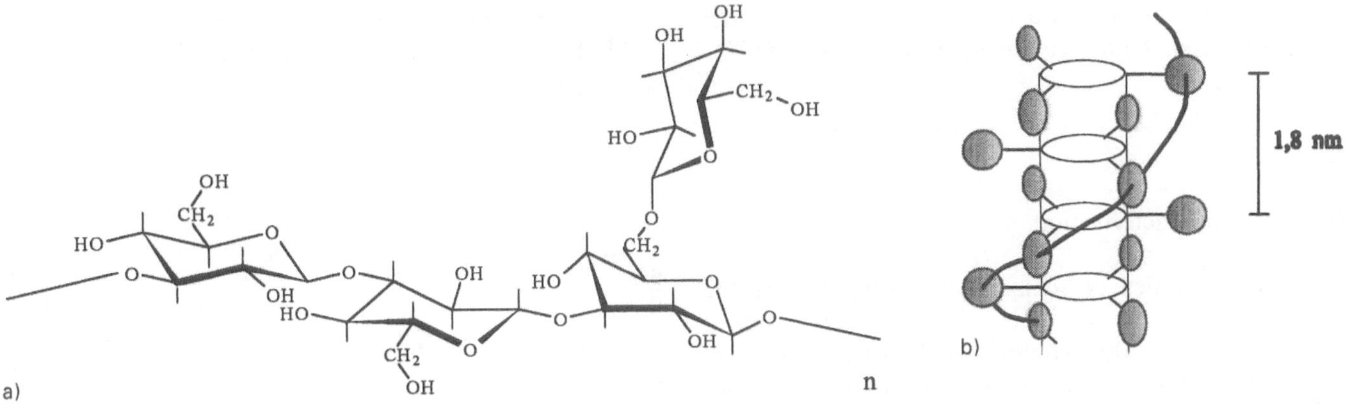

Fig. 1A Chemical structure of Schizophyllan; **B** Schematic representation of the polymer conformation

The *cis* position of the adjacent hydroxyl groups of the polymer molecule has first been underlined by different authors as a necessary condition for complexation with borate ions, but Foster [15], in 1957, first mentioned the ability of the methyl-D-glucopyranoside, which can be considered as a model of Schizophyllan, to form *trans* complexes in presence of borate ions in spite of the apparently unfavourable rigid structure to such complexation.

Experimental

Material

Schizophyllan powder was kindly provided by Taito Co. (Japan), and it consists of a slightly shear degraded sample from the native form. Its average molecular weight $\overline{Mw}$, as determined by low-angle laser light scattering technique, was found to be about $2 \cdot 10^6$ g/mol.

Borax $Na_2B_4O_7$, $10H_2O$ and sodium chloride, all of analytical grade, were from Prolabo, and pH adjustments were accomplished using a concentrated aqueous solution of sodium hydroxide.

Preparation of polymer solutions

Schizophyllan powder was gently stirred in pure water for a few hours until total dissolution. Sodium chloride was then added to get the desired salt concentration in the solution, and sodium azide was added (0.4 grams per litre) as a bactericide to prevent any biopolymer degradation.

No further purification was necessary for solutions used for gel preparation. For physico–chemical characterization of the polymer dilute solutions were successively filtered through 8, 3, 1.2 and 0.45 μm porosity filters.

Preparation of gels

Concentrated polymer solution and crosslinker solution were prepared separately at given concentrations, at desired salt level and the pH value was adjusted at the desired value using a concentrated sodium hydroxide solution. Both the polymer and the crosslinker solutions were prepared at a fixed temperature for 30 min using a temperature controller bath, and then mixed under vigorous stirring to prevent precipitation that may be caused by local over-concentration, in order to give the final desired concentration of both components.

This method allowed to study kinetics of gelation, phase diagrams and rheological behaviour of the physical gels.

Apparatus

The weight-average molecular weight, Mw, of the biopolymer sample was determined by LALLS using a Chromatix KMX-6. Shear viscosity measurements were performed with a Low-Shear 30 Contraves Rheometer operating over a range of the shear rates between 0.017 and 128 s^{-1}. The dynamic measurements were performed using a Carri-Med CSL-100 controlled stress rheometer with a cone and plate device (radius 4 cm, cone angle 2°). The apparatus was equipped with a solvent trap that prevents any dehydration phenomenon during measurements, and temperature was controlled by Peltier effect allowing an accuracy of about 0.1 °C. Total sample amount required is about 2 ml. The kinetic of gel formation was followed by measuring the storage and loss moduli G' and G'' at a fixed frequency (1 Hertz) under a low stress (0.1 N/m^2) to avoid any sample disruption over the reaction of gelation. As the sample is initially fluid before the gelation occurs a double concentric cylinder system was used. All measurements were made at 25 °C.

Results

Rheological characterization

The flow properties of schizophyllan solutions in the presence of different amounts of borax were studied: Figure 2 represents the evolution of the shear rate when increasing the borax amount for a given polymer concentration (4 g/l). It clearly appears that the rheological properties of Schizophyllan aqueous solutions are largely affected by addition of small quantities of borate ions. This phenomenon is observed as far as the polymer solution is in the semi-dilute regime. The entanglement of polymeric chains gives rise to the establishment of intermolecular physical bonds: as the quantity of borate ions becomes sufficient, it leads to the formation of physical gels. Furthermore, the higher the borax concentration in solution, the stronger the gel becomes as evidenced by creep measurements shown in Fig. 3. Relative scale for the compliance curves was used for a best representation of the elasticity recovery of the samples. As illustrated in Fig. 4, the maximum compliance strongly decreases upon increasing the borax concentration.

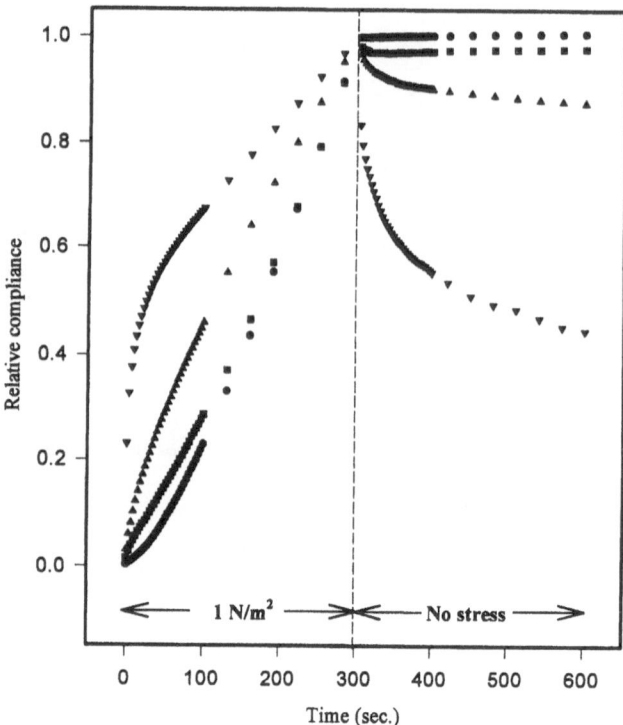

Fig. 3 Relative compliance curves as a function of borax concentration in the schizophyllan solutions during creep experiments. $Cp = 4$ g/l − NaCl 0.5 M − $T = 25$ °C − pH = 9. ● 4 mM − ■ 10 mM − ▲ 12 mM − ▼ 14 mM

Fig. 2 Influence of borax addition on the flow curves of schizophyllan solutions. $Cp = 4$ g/l − NaCl 0.5 M − $T = 25$ °C − pH = 9. From top to bottom: polymer only, 2.5, 4, 5, 6.5, 8, 10, 12 and 14 mM borax

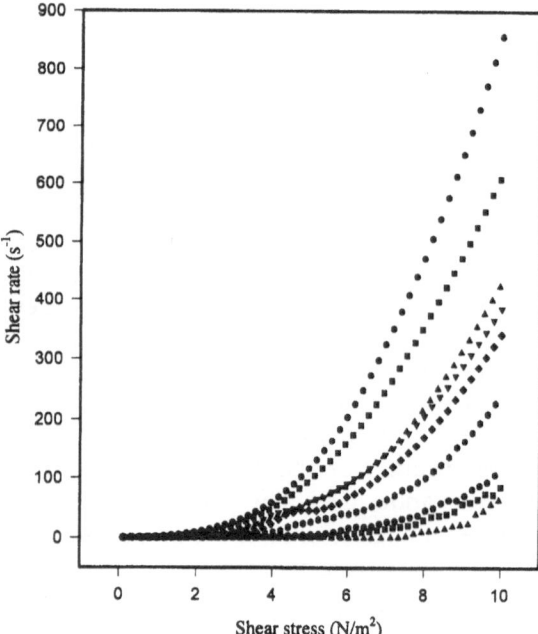

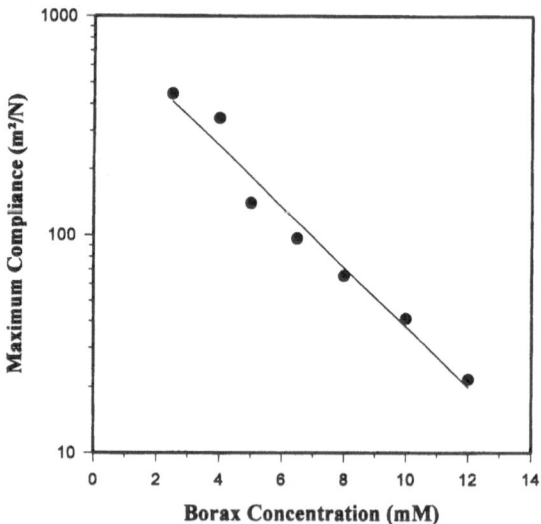

Fig. 4 Evolution of the maximum compliance during creep experiments as a function of borax content. $Cp = 4$ g/l − NaCl 0.5 M − $T = 25$ °C − pH = 9

Figure 5 shows the frequency dependence of the storage and the loss moduli, namely $G'(\omega)$ and $G''(\omega)$, for a schizophyllan−borax mixture before and after the gelation occurs. Experimental conditions were chosen in both

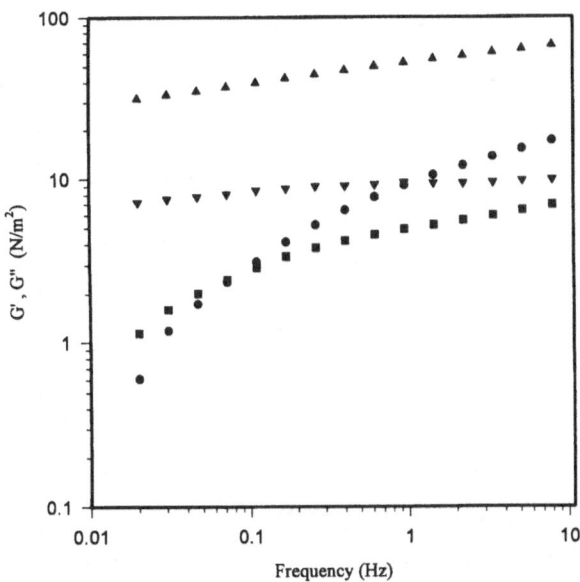

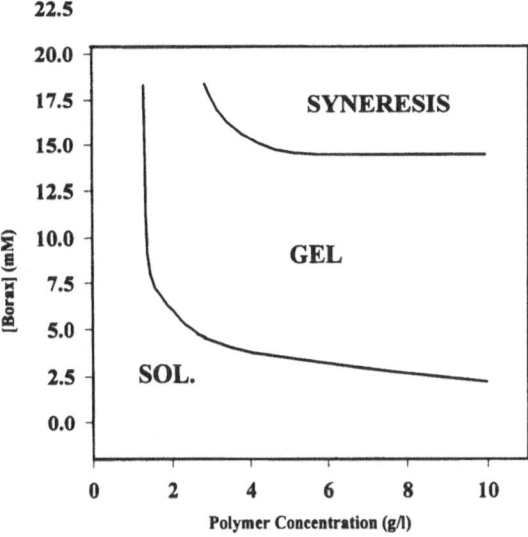

Fig. 5 Evolution of the viscoelastic parameters during the gelification reaction. $Cp = 6$ g/l – [Borax] = 14 mM – NaCl 0.75 M – $T = 25\,°C$ – pH = 9. ●, ■ G', G'' (initial) and ▲, ▼ G' and G'' (final)

Fig. 6 Phase diagram for schizophyllan – borax mixtures. NaCl 0.5 M – $T = 25\,°C$ – pH = 9

cases in order to get reasonable strain ($<10\%$) at a fixed stress over the whole frequency range. Initial measurement was carried out just after mixing the solutions, and final data were taken after gel formation, in practice at time $t = 3\,t_{gel}$. This experiment demonstrates the apparition of a marked elastic behaviour related to a weak physical gel when compared to the initial state.

Phase diagram

The gels were prepared following the procedure described in the experimental section. The mixtures were then left for a few days (at least a period time three times the gelation time) afterwhich low shear measurements ($0.08\,s^{-1}$) were made for determining the sol-to-gel transition. Visual observation of the mixtures allowed to evidence syneresis phenomenon (phase separation) occurring as the result of gel contraction and liquid expulsion from the structure due to the ⟨excess⟩ of junction points in the network.

Figure 6 represents a typical example of phase diagram for the system; three different domains can be distinguished depending on the polymer and crosslinker concentrations: sol, gel and syneresis.

The gel area appears quite narrow as a consequence of the phase separation occurring early as the borax amount increases, this phenomenon corresponds to a change in the gel state. Finally, syneresis is practically independent of polymer concentration.

Other works demonstrated that the phase diagram is highly dependent on physical parameters such as temperature, pH and salinity conditions.

Kinetic studies

The kinetic of gelation can be monitored by measuring the time dependence of $G'(\omega)$ and $G''(\omega)$ at a constant excitation frequency. As illustrated by Fig. 7, the gelation process shows two distinct time dependencies. First, the sample's rheological parameters appear quite constant, afterwhich the elastic modulus raises very sharply, therefore evidencing the onset of gelation. The critical point at which the contribution of the elastic component $G'(\omega)$ becomes predominant indicates that the macromolecules are not independent anymore as they form a network preventing individual motion. This critical point, which is considered as the gel point, is shifted towards shorter time as the crosslinker concentration is higher (Fig. 8) as previously mentioned by Han [16]. Another very interesting observation is that the higher the salt content the sooner the gelation takes place as evidenced by Fig. 9. This observation argues that the salt is of prime importance in the formation of the complexes between the borate ions and the hydroxyl sites of the polymer chain. Actually, the gelation is not observed without adding salt to the mixture, which is easily understood by considering that during complexation a negative charge has been introduced on the initially neutral polymer chain. The resulting electrostatic interactions make difficult gelation in pure water, whereas they are screened upon addition of salt, therefore favouring gelation.

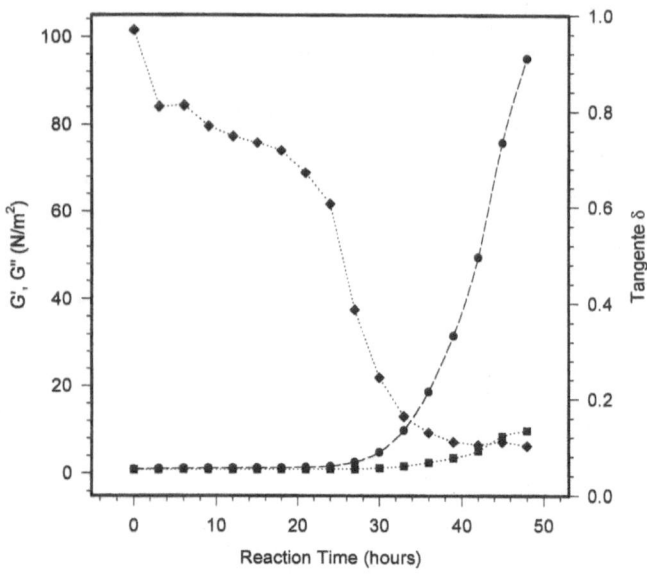

Fig. 7 Following of the gelation of schizophyllan – borax system. $Cp = 3$ g/l – [Borax] $= 12.5$ mM – NaCl 0.5 M – $T = 25\,°C$ – pH $= 9$. ● G' – ■ G'' – ◆ Tangente δ

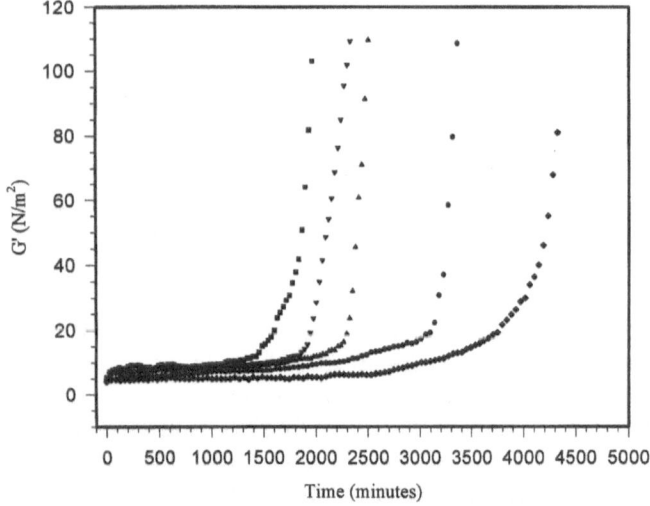

Fig. 8 Kinetics of gelation for schizophyllan in presence of borate ions. Complexing agent concentration effect. ◆ 2.5 mM ● 5 mM ▲ 8 mM ▼ 10 mM ■ 14 mM

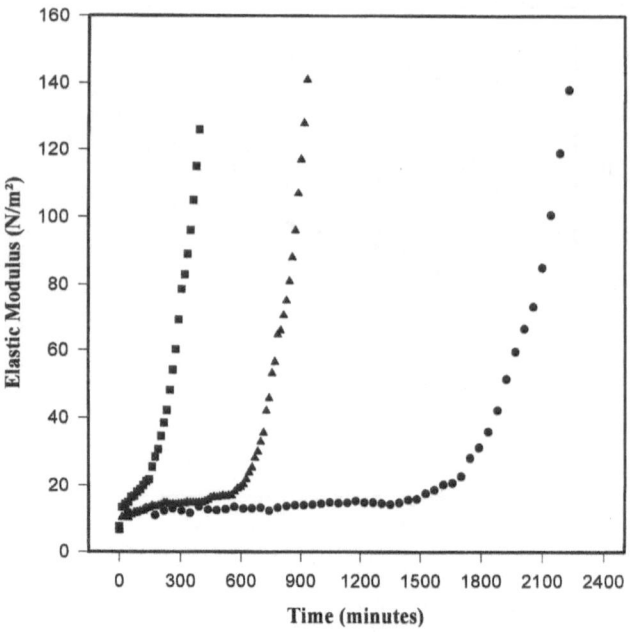

Fig. 9 Salinity influence on the kinetic of gelation. $Cp = 6$ g/l – [Borax] $= 14$ mM – $T = 25\,°C$ – pH $= 9$. ● 0.5 M – ▲ 1 M – ■ 1.5 M

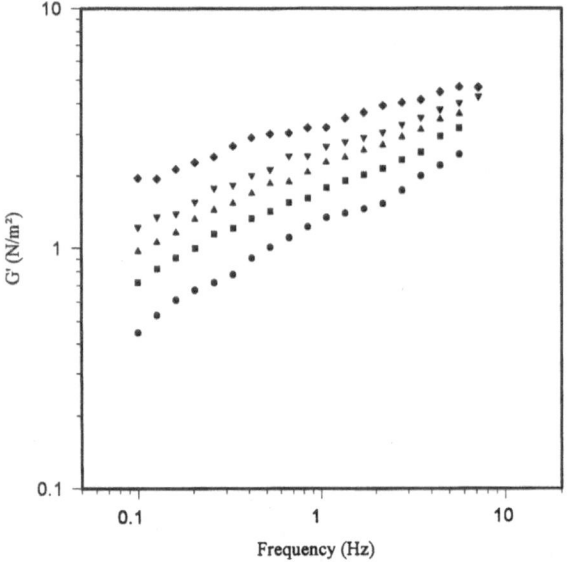

Fig. 10 Salinity influence over the viscoelastic properties of the system. $Cp = 3$ g/l – [Borax] $= 10$ mM – $T = 25\,°C$ – pH $= 9$ [NaCl]: ● 0.25 M, ■ 0.5 M, ▲ 1 M, ▼ 1.5 M, ◆ 2 M

Viscoelastic properties

As already discussed above, the gels properties are highly influenced by the concentration of both the polymer and the crosslinker. In addition, we have studied the salinity and the pH dependence of the physical properties of the gels using oscillatory experiments.

Salinity influence

A series of mixtures were prepared which contain the same amounts of polysaccharide and of crosslinker, but various amounts of salt (0.25, 0.5, 0.75, 1 and 2 mol/l respectively). Figure 10 clearly indicates that the gel properties of

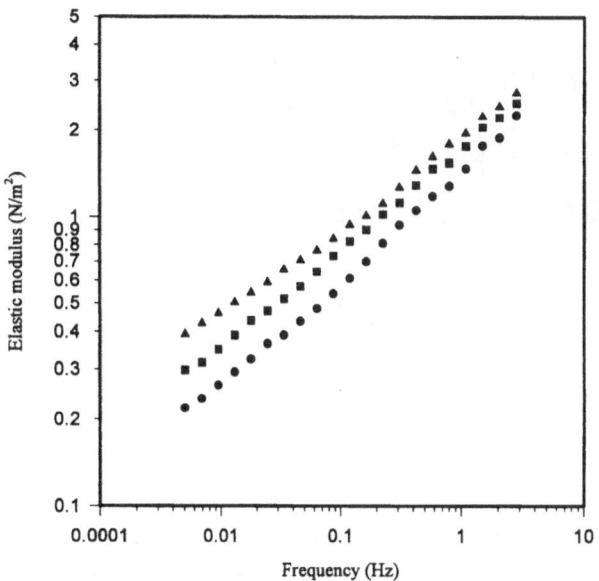

Fig. 11 pH effect on the rheological properties. $Cp = 3$ g/l – [Borax] $= 7$ mM – NaCl 0.5 M – $T = 25\,°C$. ● pH $= 9$ – ■ pH $= 10$ – ▲ pH $= 11$

pH effect

In aqueous media borax dissociates into boric acid $B(OH)_3$ and monoborate molecules $B(OH)_4^-$ (Eq. (1)), both species being in equal quantities at pH $=$ pKa ($= 9$). Most of our experiments were performed in self-buffered conditions by the presence of small amounts of borax; however it was interesting to look at the effect of changing the pH on the properties of the gels.

$$Na_2B_4O_7 + 7H_2O \rightarrow 2Na^+ + 2B(OH)_3 + 2B(OH)_4^- \quad (1)$$

$$B(OH)_3 + 2H_2O \rightleftarrows B(OH)_4^- + H_3O^+ \quad pKa \# 9 . \quad (2)$$

Figure 11 shows that the elasticity of the system is enhanced by increasing the pH, thus indicating that the number of efficient bonds becomes higher.

This can be easily explained as the ratio $[B(OH)_3]/[B(OH)_4^-]$ is highly dependent on the pH conditions: upon increasing the pH, the above equilibrium (2) is shifted to the right and the concentration of $B(OH)_4^-$ is higher.

mixtures are in close relation with the salt content of the solutions: the elastic modulus $G'(\omega)$ increases and becomes less frequency dependent with increasing the salinity. The salt not only increases the rate of gelation as demonstrated above, but it also reinforces the elastically active network junctions. At the moment this can be interpreted by considering that the effective links are stabilized by introduced ions and that there are even more physical links involved in the network.

Further work is in progress to more precisely elucidate the observed behaviour.

Conclusion

This study shows that schizophyllan can form physical gels as a result of complexation of sodium borate with hydroxyl groups favourably located on the polysaccharide chains. Many factors have been found to affect both the gelation kinetics and the overall properties of gels as concentration of interacting species, pH and salinity.

Useful information on the chemistry and the structure of such systems are expected from NMR and microscopy experiments.

References

1. Schultz RK, Myers RR (1969) Macromolecules 2(3):281
2. Keita G, Ricard A (1990) Polymer Bull 24:627
3. Osaki K, Inoue T, Ahn KH (1994) J Non-Newtonian Fluid Mech 54:109
4. Noble O, Taravel FR (1987) Carbohydr Res 166:1
5. Pezron E, Ricard A, Lafuma F, Audebert R (1988) Macromolecules 21:1121
6. Boeseken J, Van Giffen J (1920) Recl Trav Chim 39:183
7. Hermans PH (1925) Anorg Allg Chem 142:83
8. Kikomoto S, Miyajima T, Yoshizumi S, Fujimoto S, Kimura K (1970) J Agr Chem Soc Jpn 44:337
9. Tabata K, Ito W, Kojima T, Kawabata S, Misaki A (1981) Carbohydr Res 89:121
10. Takahashi Y, Kobatake T, Suzuki H (1984) Rep Prog Polym Phys Jpn 27:767
11. Singh PP, Whistler RL, Tokuzen R, Nakahara W (1974) Carbohydr Res 37:245
12. Norisuye T (1985) Makromol Chem Suppl 14:105
13. Alhaique F, Riccieri FM, Santucci E, Crezcenzi V, Gamini A (1985) J Pharm Pharmacol 37:310
14. Romanelli L, Alhaique F, Riccieri FM, Santucci E, Valeri P (1993) Pharmacological Research 27(1):127
15. Foster AB (1957) Adv Carbohydr Chem 12:81
16. Han M (1992) Thesis, Université de Rouen, France

Progr Colloid Polym Sci (1996) 102:38–41
© Steinkopff Verlag 1996

R.-H. Mikelsaar

Molecular modelling of cellulose and hyaluronan three-dimensional structure

Dr. R.-H. Mikelsaar (✉)
Molecular Modelling Laboratory
Institute of General and Molecular
Pathology
Tartu University
Veski Street 34
Tartu EE2400, Estonia

Abstract Tartu plastic space-filling atomic-molecular models and packing energy calculations were used to investigate cellulose and hyaluronan three-dimensional (3-D) structure. The molecular modelling data allowed us to propose for both biopolymers novel antiparallel structures suitable for explanation functional properties of these wide-spread molecules.

Key words Cellulose – hyaluronan – polysaccharide 3-D structure – molecular modelling

Introduction

Cellulose (CE) and hyaluronan (HA) are the most wide-spread polysaccharides in animal and plant tissues. The basic primary structure of these molecules is well known, but the three-dimensional secondary and tertiary structure as well as their functional role are not yet deciphered in details. By NMR investigations it was elucidated that CE is a mixture of two crystalline phases (α and β) whose proportion depends on the source of cellulose [1]. CE α phase is metastable and can be converted readily into the thermodynamically stable β phase [2]. Sugiyama et al. studied algal CE by electron diffraction and found that α phase has a one-chain triclinic structure and β phase is characterized by two-chain unit cell and a monoclinic structure [3]. NMR demonstrated in HA structure a two-fold helix with an extended H-bonded system and water bridges as well as extensive hydrophobic patches distributed on alternate sides of the tapelike molecule [4, 5].

The aim of the current paper was, using molecular modelling method and packing energy calculations: 1) to elucidate the most favourable stereochemical structures corresponding to unit cell parameters found in CE α and β crystalline phases, and 2) to examine possibilities of hydrophilic and hydrophobic interactions between HA molecules, to gain insights into its behaviour in tissues.

Materials and methods

Molecular modelling was carried out with Tartu plastic space-filling models having improved parameters and design [6, 7]. Packing energy calculations were carried out by a rigid-ring method [8].

Results

Molecular modelling of CE 3-D structure

Modelling with Tartu devices [9] showed that there are one parallel molecular model (P1) and two antiparallel models (A1a and A1b) fitting with CE α phase unit cell parameters. Packing energy calculations revealed that these models have following potential energy values: P1 -19.5, A1a -16.4 and A1b -15.5 kcal/mol. The same method of molecular modelling also indicated that there are one parallel (P2) and three antiparallel models (A2, A3a, A3b) fitting with CE β phase unit cell parameters.

Progr Colloid Polym Sci (1996) 102:38–41
© Steinkopff Verlag 1996

Fig. 1 Molecular structure of cellulose α phase: antiparallel chains, intra- and intermolecular hydrogen bonds. Small circles: C-atoms; large circles: O-atoms; G: glycose residue

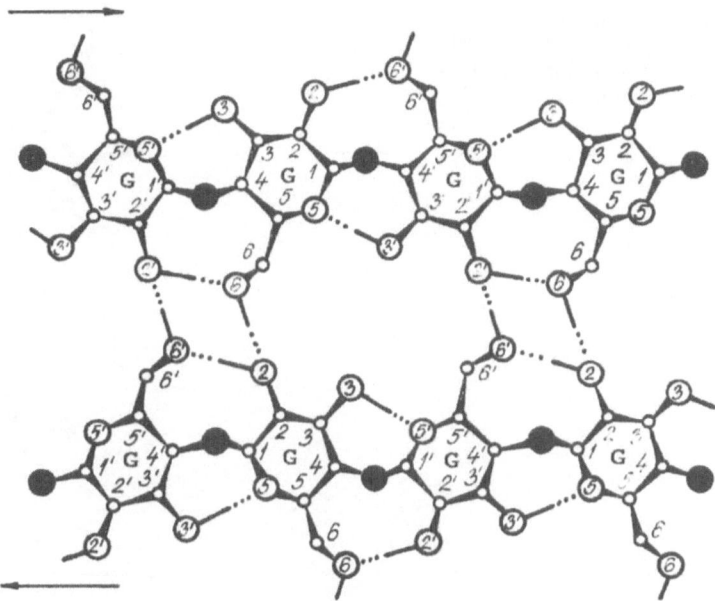

These models have potential energy: P2 −19.9, A2 −15.8, A3a −21.0 and A3b −15.1 kcal/mol.

Comparing CE parallel model P1 with antiparallel models A1a and A1b, one can see that these models are sterically very similar. Since the H-bonding network and chain polarity are not clearly established, the structure characterized by eight-chain unit cell can be considered as structure of one-chain unit cell. We think that antiparallel model A1a having lower energy (−16.4 kcal/mol) in comparison with the model A1b (−15.5 kcal/mol) is a possible candidate for a "masked" eight-chain unit cell in the CE α crystal structure described by Sugiyama et al. [3]. Probably the experimental procedure of electron diffraction in the above-mentioned investigation allowed to record only the parameters of subunits of eight-chain unit cells. The antiparallel structure is not excluded also for cellulose β crystalline phase because the same authors marked that the exact position of the cellulose chains in the monoclinic unit cell was even more difficult to be found as long as the structure refinement was not achieved. No doubt, A3a is the best antiparallel model for fitting to the unit cell structure of β crystalline phase because it has an excellent packing energy: −21.0 kcal/mol.

General chemical formulae and photographs of antiparallel models A1a and A3a are demonstrated in Figs. 1–3.

Molecular modelling of HA 3-D structure

Molecular modelling was carried out for HA chains with and without water bridges [10]. It was elucidated that

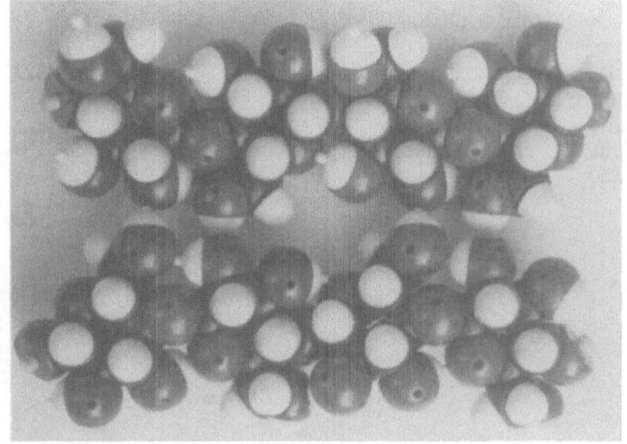

Fig. 2 Molecular model of cellulose α phase structure (cf. Fig. 1)

there are two sterically possible different secondary structures for HA chains lacking water bridges and four possible structures for chains with water bridges. Our models revealed that hydrophobic contacts are possible only between HA chains lacking water bridges in the secondary structure.

Observation of the peculiarities of the HA secondary structure reveals that most polar groups can form intramolecular H-bonds. However, two groups are "free": hydroxymethyl and an oxygen atom of the carboxylate group. These groups could mediate intermolecular hydrophilic interactions in assemblies containing large numbers of HA molecules. Molecular modelling shows that H-bonds between hydroxymethyl and carboxylate groups

Fig. 3 Molecular structure of cellulose β phase: antiparallel chains, intra- and intermolecular hydrogen bonds

Fig. 4 Molecular structure of hyaluronan: hydrogen bonds between hydroxymethyl and carboxylate groups of antiparallel chains. Small circles: C-atoms; large circles: O- and N-atoms; G: glucuronic acid residue; N: acetamidoglucose residue

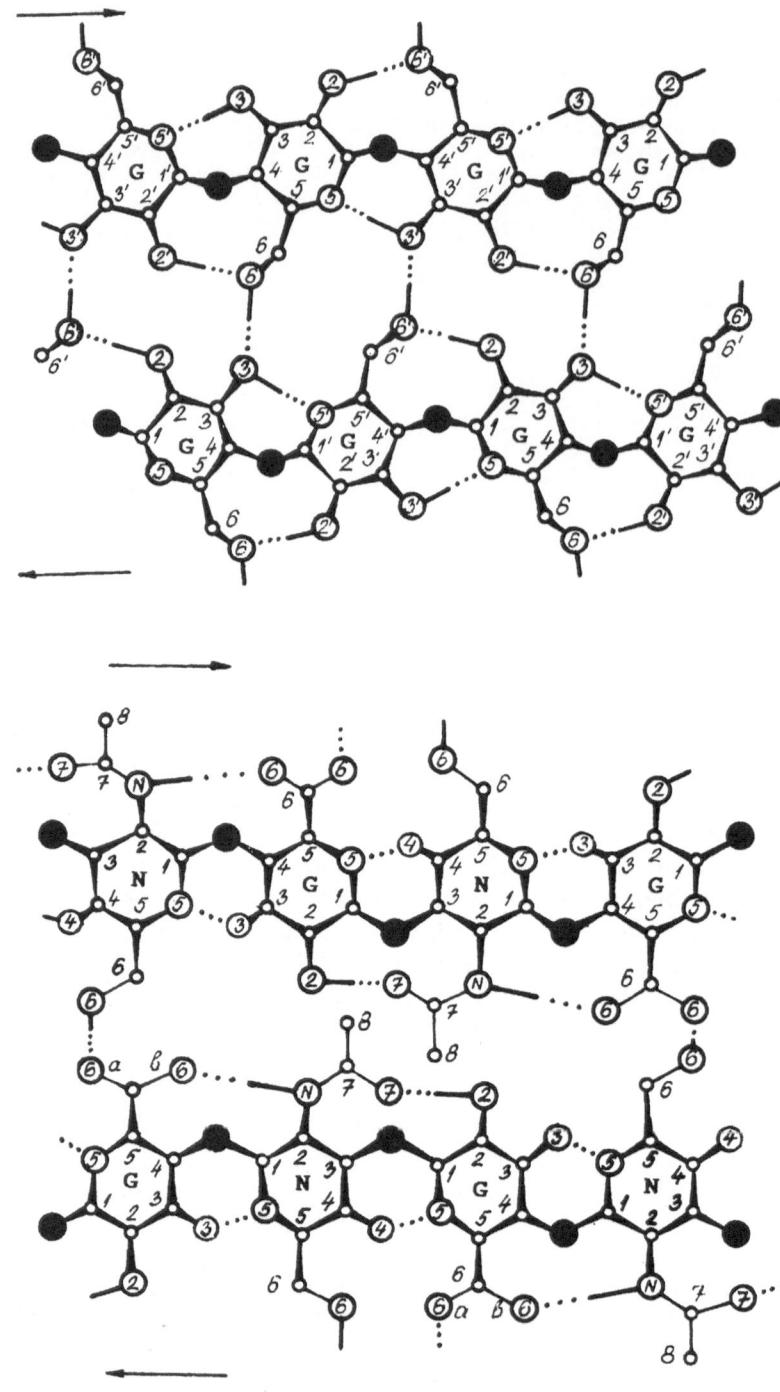

are possible only between antiparallel HA molecules (Figs. 4 and 5). Each disaccharide residue can form two H-bonds, so that bonds on one side of the HA molecule alternate with analogous bonds on the other side. Such H-bonding can join antiparallel HA molecules into sheets which are planar or curved. The latter may form tubular structures that are closed along all circumferences, or are open to one side.

Discussion

Our investigation revealed that both parallel and antiparallel CE structure models can be proposed, based on α and β crystalline phase unit cell parameters. Sugiyama et al. [3] interpret their data in the light of all-parallel-structure view. However, the all-parallel-structure conception meets

Progr Colloid Polym Sci (1996) 102:38–41
© Steinkopff Verlag 1996

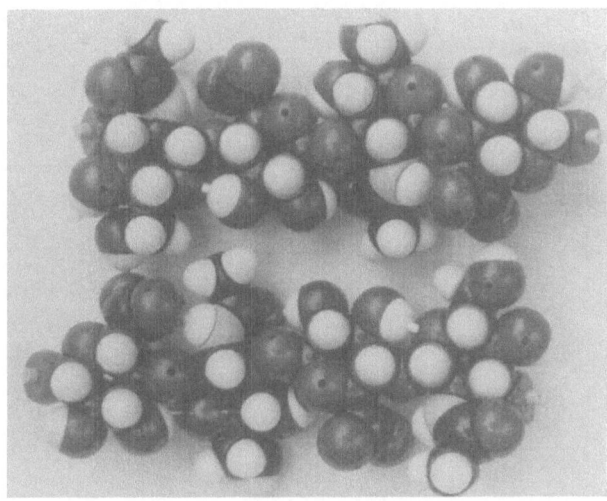

Fig. 5 Molecular model of hyaluronan structure (cf. Fig. 4)

many difficulties when one tries to explain the molecular mechanisms of CE phases interconversion. If the antiparallel models A1a and A2a are thought to be the candidates for structures of CE α and β crystalline phases, correspondingly, it would be easy to explain the experimental data on the interconversion of these phases. Model A1a, having relatively high energy -16.4 kcal/mol, is metastable, and during annealing process can easily be converted irreversibly into the model A3a, having very low energy -21.0 kcal/mol. Because the models A1a and A3a have different H-bonding network, our conception on antiparallel cellulose structure is also in accordance with Raman-spectroscopic data [11]. The high energy barrier between α and β crystalline phases, bound with rearrange-

ment of hydrogen bonds, may be the reason why the proportion of these phases in different samples of native cellulose is relatively constant. The order of chain rearrangements in the transformation of models A1a → A3a, probably corresponding to the cellulose $\alpha \to \beta$ phase conversion, seems to be simple. After the interruption in model A1a of intrasheet bonds O2–H $\cdots$ O6 the alternate sheets of parallel chains must move one step diagonally in the plane of these sheets and new intrasheet bonds O3–H $\cdots$ O6 characteristic of model A3a should be formed.

Molecular modelling by Tartu space-filling atomic models confirmed that H-bonds are present in the HA secondary structure, reducing configurational flexibility and increasing stiffness of this molecule [12]. The main finding was that of H-bonding between antiparallel HA molecules. These bonds are similar to the polar interactions between hydroxymethyl and glycol groups of CE chains and between hydroxymethyl and acetamido groups of chitin molecules. By contrast with CE, where intermolecular H-bonds are formed in the course of biosynthesis, HA H-bonded sheets may form at high concentrations as a self-aggregation process. In the process of self-aggregation polar interactions between HA chains are apparently supplemented by hydrophobic contacts. The combination of hydrophilic and hydrophobic interactions gives highly-ordered structures which may be responsible for the unique visco-elastic and physiological properties of HA in tissues. Self-aggregated planar sheets may give rise to filaments of varied thickness and sheets of material, as seen by electron microscopy. H-bonds between crossed sheets could serve as a basis for extensive branched networks. Curved sheets of HA molecules might form channels through which e.g. water-soluble molecules could diffuse.

References

1. Atalla RH, VanderHart DL (1984) Science 223:283–285
2. Horii F, Yamamoto H, Kitamaru R, Tanahashi M, Higuchi T (1987) Macromolecules 20:2946–2949
3. Sugiyama J, Vuong R, Chanzy H (1991) Macromolecules 24:4168–4175
4. Scott JE, Cummings C, Brass A, Chen Y (1991) Biochem J 274:699–705
5. Scott JE, Cummings C, Greiling H, Stuhlsatz HW, Gregory JD, Damle SP (1990) Int J Biol Macromol 12:180–184
6. Mikelsaar R-HN, Bruskov VI, Poltev VI (1985) New precision spacefilling atomic–molecular models, Scientific Center of Biological Research of the Academy of Sciences of the USSR in Pushchino, pp 1–44
7. Mikelsaar R (1986) Trends Biotechnology 6:162–163
8. Pertsin AJ, Nugmanov OK, Marchenko GN, Kitaigorodsky AI (1984) Polymer 25:107–114
9. Mikelsaar R-H, Aabloo A (1995) Mol Mat 5:165–173
10. Mikelsaar R-H, Scott JE (1994) Glycoconjugate Journal 11:65–71
11. Atalla RH (1989) In: Kennedy JF, Phillips GO, Williams PA (eds) Cellulose structural and functional aspects. Ellis Horwood Ltd, New York, Chichester, Brisbane, Toronto, pp 61–73
12. Scott JE, Heatley F, Hull WE (1984) Biochem J 220:197–205

Progr Colloid Polym Sci (1996) 102:42–46
© Steinkopff Verlag 1996

T.F. Irzhak
N.I. Peregudov
M.L. Tai
V.I. Irzhak

Bond blocks concept in kinetics of formation and destruction of polymer gels

T.F. Irzhak · N.I. Peregudov
Dr. V.I. Irzhak (✉)
Institute of Chemical Physics
Chernogolovka
Moscow distr., 142432, Russia

M.L. Tai
Mathematical Department
Nizhnii Novgorod State University

Abstract Bond blocks concept is used for kinetic description of the polymer formation processes and their structures including systems with reversible reactions of polycondensation or destruction of polymer chains. It is shown that the oscillatory type MMD can occur. The kinetic conditions of its occurrence are obtained.

Key words Bond blocks – polycondensation – copolycondensation – molar mass distribution – substitution effect

Introduction

Bond blocks concept developed some years ago [1, 2] is useful for description of complicated polymers structure and processes of their formation. Bond blocks of any structure can be represented as a connected graph **G** independent of the macromolecule structure in which it is located. Edges and vertexes correspond to bonds and units of the polymer. Concentration of **G** is denoted by $y\{G\}$. The complete set of concentrations of all **G** characterizes composition and structure of the system unambiguously [1, 3].

To describe the process of polymer formation, we have to write the system of differential equations. The rules of solving this problem are formulated in [1–6].

The major results of the works can be summarized as follows:

– the kinetics of a polymer formation, molecular and topological characteristics of the polymer formed can be described using the acting mass law in terms of changeable bond block concentrations;

– the systems of differential equations describing the kinetics of the process are infinite but complete; this means that one can cut them and solve numerically with any desired degree of exactness;

– for description of the polymer formation kinetics and their molecular and topological structures, one can use a set of bond blocks.

Here, we are going to describe the reversible reactions of polycondensation and destruction of polymer chains using the bond blocks concept.

End bond blocks

The key to formulation of the system of differential equations lies in generalizing the idea of end groups. Actually, if we deal with chemical bonds (the simplest case of bond blocks), we consider only one or two end units:

$$\sim a + \sim a \rightarrow \sim a\text{--}a \sim \,,$$

$$\sim a + \sim b \rightarrow \sim a\text{--}b \sim \,,$$

$$\sim b + \sim b \rightarrow \sim b\text{--}b \sim \,,$$

where $\sim a$ and $\sim b$ denote end units "a" and "b", $\sim a\text{--}a \sim$, $\sim a\text{--}b \sim$ and $\sim b\text{--}b \sim$ are the bonds between these units. This approach is widely used since Flory's works [7] on polymer formation kinetics. In case of more complicated bond blocks, we have to use more complicated end blocks. For instance, the block $\sim abbb \sim$

Progr Colloid Polym Sci (1996) 102:42–46
© Steinkopff Verlag 1996

is formed by reactions:

$$\left.\begin{array}{l} \sim a + \ \sim bbb \\ \sim ab + \ \sim bb \\ \sim abb + \ \sim b \end{array}\right\} \to \ \sim abbb \sim \ .$$

Therefore, it is necessary to have a system of equations for each end blocks. It is obvious that these systems are complete and there are no difficulties to apply any numeric method to calculate the concentrations of each end block.

As an example of using this approach, let us consider the system consisting of branching three-functional units "a" and bifunctional units "b". The reaction is supposed to occur between "a" and "b", "b" and "b"; reaction between "a" and "a" is impossible.

$$a_3 + b_2 \to a_2 b_1$$

$$\sim a_2 + b_2 \to \ \sim a_1 b_1$$

$$\sim a_1 + b_2 \to \ \sim a_0 b_1$$

$$a_3 + \ \sim b_1 \to a_2 b_0 \sim$$

$$\sim a_2 + \ \sim b_1 \to \ \sim a_1 b_0 \sim$$

$$\sim a_1 + \ \sim b_1 \to \ \sim a_0 b_0 \sim$$

$$b_2 + b_2 \to b_1 b_1$$

$$b_2 + \ \sim b_1 \to b_1 b_0 \sim$$

$$\sim b_1 + \ \sim b_1 \to \ \sim b_0 b_0 \sim \ . \tag{1}$$

The indexes denote the number of functional groups number (maximum 3 in the case of units "a" and 2 in the case of "b"). If the index is equal to 0 it means that the units are located inside of the block. In the opposite case the units form the end block.

If the substitution effect operates, each of the listed reactions must be characterized by different kinetic constants. In this case only the numeric methods of solution are effective.

It is very easy to generalize the system (1) to describe more complicated structures.

$$\sim ba_2 + b_2 \to \ \sim ba_1 b_1$$

$$\sim ba_1 + b_2 \to \ \sim ba_0 b_1$$

$$\sim ba_2 + \ \sim b_1 \to \ \sim ba_1 b_0 \sim$$

$$\sim ba_1 + \ \sim b_1 \to \ \sim ba_0 b_0 \sim$$

$$\sim ab_1 + b_2 \to \ \sim ab_0 b_1$$

$$\sim bb_1 + \ \sim b_1 \to \ \sim bb_0 b_0 \sim \ . \tag{2}$$

Corresponding systems of differential equations have positive and negative terms if we consider the end bond blocks, and there are positive ones only in the case of "internal" bond blocks. But in any case the systems are complete.

It is obvious that for description of copolymer systems, we need to use the same approach. The difference arises due to necessity to consider many more components in the reaction systems. For example, if there are three copolymers, a, b and c, we have to consider the end blocks:

$$\sim a, \sim b, \sim c,$$

$$\sim aa, \ \sim ab, \ \sim ac, \ \sim ba, \ \sim bb, \ \sim bc, \ \sim ca, \ \sim cb, \ \sim cc,$$

$$\sim aaa, \ \sim aab, \ \sim aac, \ \sim aba \text{ and so on.}$$

As it follows from the meaning of the bond block definitions,

$$[\sim a] = [\sim aa] + [\sim ba] + [ca],$$

$$[\sim aa] = [\sim aaa] + [\sim baa] + [\sim caa] \text{ and so on.}$$

Here, $[x]$ denotes concentration of component "x".

For the copolymer systems the bond blocks approach is very convenient and in the case of many comonomer units is unique. Actually, the number of the possible chains of the different copolymer structure $N(m, n) = m^n$, where m is the number of the comonomer units types, n is the chain length. For $m = 2$ and $n = 100$ $N(2, 100) = 10^{30}$, i.e., much more than Avogadro number N_A. The number of reacting species of each sort is not enough for application of the acting mass law due to its statistical nature. If $m > 2$, for example, $m = 10$, $n \geq 24$ leads to the same situation: $N(10, 24) > N_A$. Therefore, it is impossible to consider polymer chains of definite structure as kinetic components. However, bond blocks may be considered in that role. Note, that Lifshits paid attention to this problem about 10 years ago [8].

The more complicated the bond block considered, the greater the number and the more complicated the end block taken into account. However, in any case the systems of equations are complete.

Destruction reactions

Reactions of destruction of polymer chains generally result in the incomplete system of differential equations. However, if there are no substitution effects for the kinetic constants of chain destruction or back reactions, the system of differential equations is complete. Below, the reversible linear polycondensation is considered as a simplest example.

In this case the scheme can be written as follows:

$$X_1 + X_1 \xleftarrow{\ \frac{k_1}{k_1^1}\ } X_2 ,$$

$$X_1 + X_i \xleftarrow{\ \frac{k_2}{k_2^1}\ } X_{i+1}, \quad i = 2, 3, \dots$$

$$X_i + X_j \xleftarrow{\ \frac{k_3}{k_3^1}\ } X_{i+j}, \quad i, j = 2, 3, \dots \cdot \tag{3}$$

Corresponding system of differential equations for the chain concentration x_i is:

$$dx_1/dt = -k_1 \cdot x_1^2 - k_2 \cdot x_1 \cdot \sum x_j + k_1^1 \cdot x_2 + k_2^1 \cdot \sum x_j$$

$$dx_2/dt = 1/2 \cdot k_1 \cdot x_1^2 - k_2 \cdot x_1 \cdot x_2 - k_3 \cdot x_2 \cdot \sum x_j$$
$$- 1/2 \cdot k_1^1 \cdot x_2 + k_2^1 \cdot x_3 + k_3^1 \cdot \sum x_j$$

$$dx_3/dt = k_2 \cdot x_1 \cdot x_2 - k_2 \cdot x_1 \cdot x_3 - k_3 \cdot x_3 \cdot \sum x_j - k_2^1 \cdot x_3$$
$$+ k_2^1 \cdot x_4 + k_3^1 \cdot \sum x_j \dots$$

$$dx_i/dt = k_2 \cdot x_1 \cdot x_{i-1} - k_2 \cdot x_1 \cdot x_i - k_3 \cdot x_i \cdot \sum x_j$$
$$+ k_3/2 \cdot \sum x_1 \cdot x_{i-1} - k_2^1 \cdot x_i - k_3^1 \cdot (i-3)/2 \cdot x_i$$
$$+ k_2^1 \cdot x_{i+1} + k_3^1 \cdot \sum x_j \tag{4}$$

for $i = 4, 5, \dots$

At $t = 0$ $x_1 = x_{10} = 1$, $x_i = 0$ for $i = 2, 3, \dots$

For the bond block concentrations y_i:

$$dy_1/dt = k_1/2 \cdot (y_0 - 2y_1 + y_2)^2 + (y_1 - y_2)$$
$$\cdot \{k_2 \cdot (y_0 - 2y_1 + y_2) + 1/2 \cdot k_3 \cdot (y_1 - y_2)\}$$
$$- 1/2 \cdot k_1^1 \cdot y_1 + (k_1^1 - k^{21}) \cdot y_2$$
$$- 1/2 \cdot (k_1^1 - 2k_2^1 + k_3^1) \cdot y_3$$

$$dy_2/dt = (y_1 - y_2) \cdot \{k_2 \cdot (y_0 - y_1) + (k_3 - k_2) \cdot (y_1 - y_2)\}$$
$$- k_2^1 \cdot y_2 - (k_3^1 - k_2^1) \cdot y_3$$

$$dy_i/dt = (y_{i-1} - y_i) \cdot \{k_2 \cdot (y_0 - y_1) + (k_3 - k_2) \cdot (y_1 - y_2)\}$$
$$+ k_3/2 \cdot \sum (y_{j-1} - y_j) \cdot (y_{i-j} - y_{i-j+1})$$
$$- (k_2^1 + (i-2)/2 \cdot k_3^1) \cdot y_i - y_{i+1} \cdot (k_3^1 - k_2^1) \tag{5}$$

for $i = 3, 4, \dots$

At $t = 0$ $y_i = 0$; $y_0 = 1$ at any time.

Relation between x and y is [1, 3]:

$$x_i = y_{i-1} - 2y_i + y_{i+1} . \tag{6}$$

As was noted above, y_i depends on y_{i+1} if $k_1^1 \neq k_2^1 \neq k_3^1$, i.e., the system is incomplete. If $k_1^1 = k_2^1 = k_3^1$ system becomes complete, i.e., y_i does not depend on y_{i+1}.

Table 1 Effect of values of kinetic constants on the calculation exactness of the chain concentration δ ($k_1 = 1$, $k_3 = 9$). $\delta = (x_{ex} - x)/x_{ex}$ where x_{ex} is the exact value of x

i	k_2	k^1	$n = 9$	$n = 19$	$n = 39$
1	1.0	1.0	0.00058	0.00000	0.00000
	1.0	2.5	0.02527	0.00044	0.00001
	0.6	1.0	0.00062	0.00000	–
	0.6	2.5	0.04153	0.00027	
	0.1	1.0	0.00162	0.00001	0.00000
	0.1	2.5	0.05343	0.00013	0.00001
4	1.0	1.0	0.02017	0.00006	0.00000
	1.0	2.5	0.15315	0.00194	0.00006
	0.6	1.0	0.01475	0.00003	–
	0.6	2.5	0.18780	0.00118	
	0.1	1.0	0.01797	0.00007	0.00000
	0.1	2.5	0.19421	0.00054	0.00005
8	1.0	1.0	2.74421	0.00058	0.00001
	1.0	2.5	5.03249	0.00496	0.00015
	0.6	1.0	4.13022	0.00039	–
	0.6	2.5	2.01598	0.00321	–
	0.1	1.0	3.69464	0.00075	0.00002
	0.1	2.5	0.63144	0.00164	0.00018

The chain concentration calculated by numeric solution of the systems (4), (5) and (6) with different number of the equations taken into account are presented in Table 1. System (5) allows to obtain the exact results at any condition of calculation.

Oscillatory molar mass distribution

Results of numerical calculation of MMD of linear homopolymer are presented in Fig. 1. As one can see, some sets of kinetic constants lead to oscillatory character of MMD (OMMD). The character of MMD did not change in the course of reaction (Fig. 2) when the reaction is irreversible. However, reversibility leads to decay of this effect with conversion (Fig. 3).

Let us find the conditions for realization of OMMD.

As one can see, the row of chain concentration in the case of OMMD is:

$x_1 > x_2 > x_3 < x_4 > x_5 < x_6$ and so on, i.e., beginning from the third member the concentration of the odd ones becomes more than concentrations of even ones. Generally, for trivial distributions the row is monotonously decreased, i.e.,

$x_1 > x_2 > x_3 > x_4 > x_5$ and so on.

Comparing the sums of the chain concentration with odd and even unit number, $z_e = \sum_1 x_{2n+1}$ and $z_0 = \sum_2 x_{2n}$, allows us to obtain the criterion of OMMD arising: $z_e > z_0$ for trivial distribution and $z_e < z_0$ for oscillatory ones.

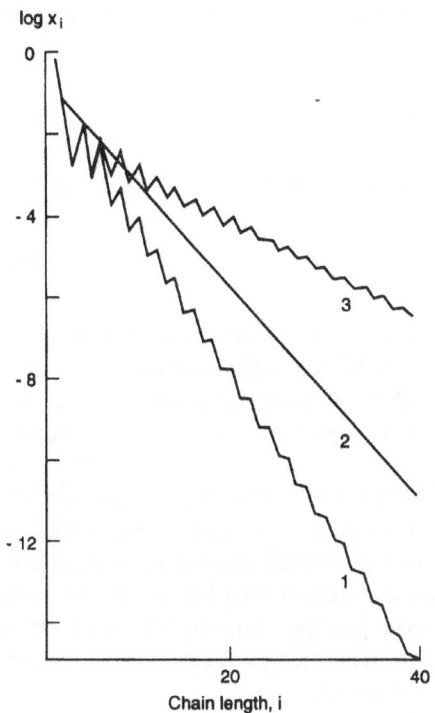

Fig. 1 Curves of molar mass distribution at irreversible polycondensation process. $k_1 = 1$; $k_2 = 0.1(1)$, $3.0(2)$ and $0.4(3)$; $k_3 = 9(1 \text{ and } 2)$, $100 (3)$ conversion is 0.2

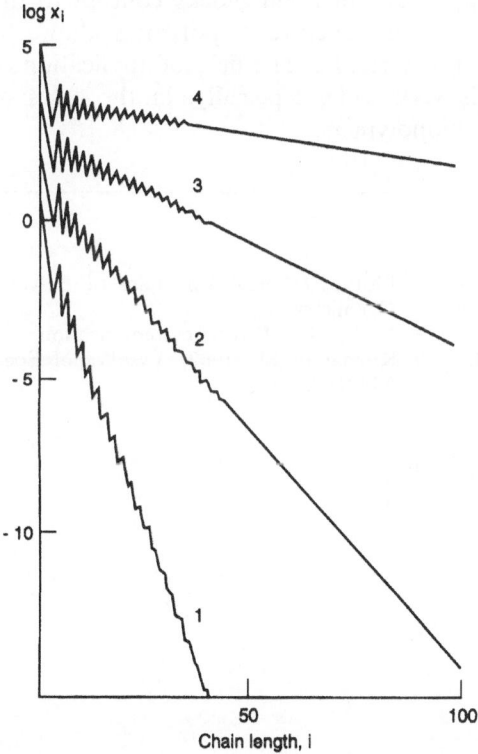

Fig. 2 Curves of molar mass distribution at irreversible polycondensation process. $k_1 = 1$; $k_2 = 0.1$; $k_3 = 10$; conversion is $0.2(1)$, $0.4(2)$, $0.6(3)$, $0.8(4)$

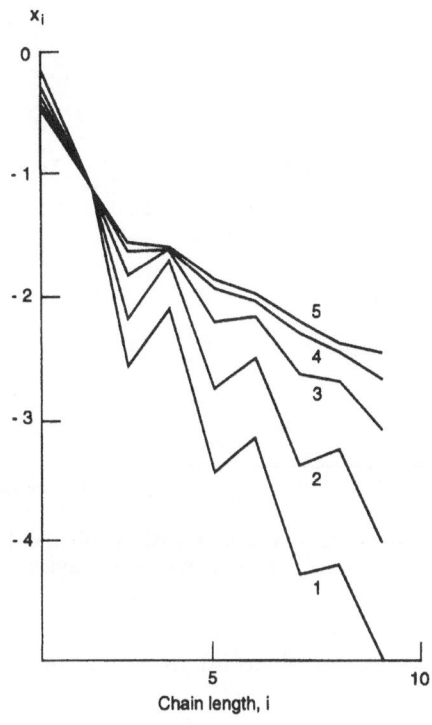

Fig. 3 Curves of molar mass distribution at reversible polycondensation process. $k_1 = 1.0$; $k_2 = 0.1$; $k_3 = 9.0$; $k' = 1.0$; ultimate conver;sion $\alpha_\infty = 0.43$. conversion α: $0.14(1)$, $0.21(2)$, $0.31(3)$, $0.39(4)$, $0.43(5)$

From (3), one can obtain the kinetic scheme and z_e and z_0:

$$X_1 + X_1 \rightarrow X_2 \quad k_1; \qquad X_2 + Z_1 \rightarrow Z_1 \quad k_3;$$

$$X_1 + X_2 \rightarrow Z_1 \quad k_2; \qquad X_2 + Z_2 \rightarrow Z_2 \quad k_3;$$

$$X_1 + Z_1 \rightarrow Z_2 \quad k_2; \qquad Z_1 + Z_1 \rightarrow Z_2 \quad k_3;$$

$$X_1 + Z_2 \rightarrow Z_1 \quad k_2; \qquad Z_1 + Z_2 \rightarrow Z_1 \quad k_3;$$

$$X_2 + X_2 \rightarrow Z_2 \quad k_3; \qquad Z_2 + Z_2 \rightarrow Z_2 \quad k_3. \qquad (7)$$

This scheme leads to the system of differential equations:

$$dx_1/dt = -k_1 x_1^2 - k_2 x_1(x_2 + z_e + z_0)$$

$$dx_2/dt = 1/2 k_1 x_1^2 - k_2 x_1 x_2 - k_3(x_2 + z_e + z_0)$$

$$dz_e/dt = k_2 x_1 x_2 - k_2 x_1 z_e + k_2 x_1 z_0 - k_3 z_e^2$$

$$dz_0/dt = 1/2 k_3 x_2^2 + k_2 x_1 z_e - k_2 x_1 z_0$$

$$+ 1/2 k_3 z_1^2 - k_3 z_1 z_2 - 1/2 k_3 z_2^2. \qquad (8)$$

46

T.F. Irzhak et al.
Bond blocks in polymer gels

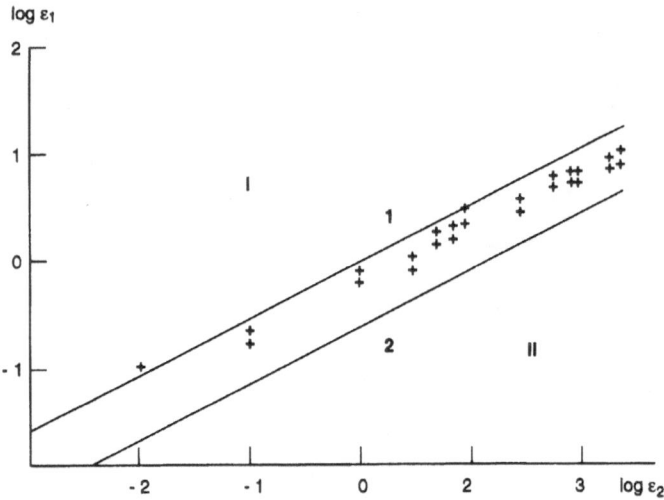

Fig. 4 Region of kinetics constant in which OMMD is observed. I-condition (10), 2-condition (12); I-no oscillation; II-oscillations are observed

Applying the steady state principle to Eq. (8), one obtains:

$$\frac{k_3(z_1 + z_2)}{k_1 x_1} < \frac{2k_2^2}{2k_1 k_3 - 3k_2^2} .$$ (9)

The necessary condition of inequality (9) to be satisfied is: $2k_1 k_3 > 3k_2^2$ or

$$\xi_1 < \frac{\sqrt{2}}{3} \xi_2 ,$$ (10)

where $\varepsilon_1 = k_2/k_1$, $\varepsilon_2 = k_3/k_1$.

This condition is shown in Fig. 4 (curve 1).

Assuming $u = (z_e + z_0)/x_1$ from (8) it is possible to obtain:

$$\frac{du}{dt} = \frac{1 + 2u + (2\xi_1 - \xi_2)u^2}{2x_1(1 + \xi_1 u)} .$$ (11)

Solution of (11) leads to condition:

$$\frac{\xi_1}{\sqrt{\xi_2}} = \frac{(\sqrt{2} - 1)}{2} .$$ (12)

This condition is also shown in Fig. 4 (curve 2). As one can see, calculated points lay between these curves.

Note, the OMMD was discovered earlier (see, for e.g., [9]). However, there was no kinetic criterion for its realization.

The problem of OMMD of polymer chain has important significance when we deal with polymer networks. In the case of alternative copolymerization processes, chains with odd or even units number display as chains with different or similar end groups. The equation of chain functionality is very important for problems of cyclization, network formation and so on.

Conclusion

Application of approach of bond blocks concept to the formation kinetics and structure of polymers allows to obtain a series of new results. The field of application of this approach is very wide, especially, in the cases of copolymers and biopolymers.

References

1. Irzhak VI, Tai ML (1983) Vysokomol Soed 25A:2305–2310
2. Irzhak VI, Tai ML (1981) Dokl Akad Nauk SSSR 259:856–859
3. Irzhak VI, Tai ML, Peregudov NI, Irzhak TF (1994) Coll Polym Sci 272:523–529
4. Irzhak TF, Peregudov NI, Tai ML, Irzhak VI (1994) Polymer Science 36A: 754–757
5. Irzhak TF, Peregudov NI, Tai ML, Irzhak VI (1995) Polymer Science 37A:446–450
6. Irzhak TF, Peregudov NI, Tai ML, Irzhak VI (1995) Vysokomol Soed 37B: 2071–2075
7. Flory PJ (1953) Principles of Polymer Chemistry
8. Lifshits IM, Private communication
9. Kuchanov SI (1982) Vysokomol Soed 24A:2179–2183

Progr Colloid Polym Sci (1996) 102:47–50
© Steinkopff Verlag 1996

N.N. Volkova
V.P. Tarasov
L.P. Smirnov
L.N. Erofeev

Sol–gel transition and molecular dynamics in the systems based on the copolymer of methyl methacrylate with methacrylic acid

Dr. N.N. Volkova (✉) · V.P. Tarasov
L.P. Smirnov · L.N. Erofeev
Institute of Chemical Physics
in Chernogolovka
RAS
Chernogolovka
Moscow region, Russia
142 432
E-mail: Nvolkova@icp.ae.ru

Abstract The NMR method is used to study sol–gel transition in the system based on the copolymer of methyl methacrylate with methacrylic acid. The NMR studies showed that sol–gel transition proceeds via three stages. At the first one, diffusion of the swelling agent causes devitrification of the copolymer; simultaneously, on the macroscopic scale the aggregates of the latex particles disintegrate. At the second stage, a gel is formed that has non-equilibrium structure. The latter is rearranged during the third stage. The ratio of the rates of the processes determining these three stages depends on the concentration of the extender. Upon completion of sol–gel transition, the system is a polyphase one. It comprises highly elastic, crystalline (or vitreous), and transitional phases. NMR spectroscopy provides a way to measure the ultimate degree of compatibility of a polymer-swelling agent pair.

Key words Sol–gel transition – compatibility – plastigel – plastisol

Experimental

We studied methyl methacrylate-methacrylic acid copolymer (MMA–MAA) containing up to 10% of the acid. $M = 1.08 \times 10^6$. The copolymer granules were high porous aggregates constituted of agglomerated spheres of 0.5 μm in diameter [1]. A plasticizer was a fluorine containing organic compound (fluorine plasticizer-FP) with $T_{melt} = 280$ K.

The measurements were performed on a multipulse Fourier NMR spectrometer (RI-2303, produced in Russia) at ^{1}H(60 MHz) and ^{19}F [2]. Free induction decay (FID) was measured by the Carr–Purcell technique [3]. The initial parts of FID were measured by "magic sandwich" [4]. The parameters of molecular motion of the studied system were calculated from temperature dependencies of spin–lattice relaxation time in rotated frame under multipulse spinlocking conditions [5, 6].

Results

Kinetics of gel formation was studied at 328 K by changes in the FID form at sol–gel transition in the system characterized by the plasticizer – polymer weight ratio "b" of 3.6 and 8. FID of the initial sample (Fig. 1, curve 1) consists of two parts different in form and decay time. The initial short Gaussian part corresponds to the vitrified and the long one to the nonvitrified polymer. The spin–spin relaxation time T_2 of the plasticizer is much longer than that of the polymer. Therefore, the form of FID of nearly 200 μs in duration depends directly on the polymer mobility, whereas the effect of the plasticizer is in the change of the polymer mobility caused by plasticization.

The relaxation times $T_2(i)$ and portions $p(i)$ of the short ($i = 1$) and long ($i = 2$) ^{1}H FID components are shown in Fig. 2. The mobility of the rigid vitrified fraction of the polymer slightly depends on "b" (Fig. 2, curve 3). At exposure at 328 K the portion of protons belonging to this fraction of the polymer decreases continuously. In other words, the temperature of the copolymer vitrification decreases with the plasticizer diffusion into the polymer.

The value of b appreciably affects the changes in mobility of the plastified fraction of the system. As indicated in Fig. 2 (curve 2), at $b = 8$ $T_2(2)$ first increases and then

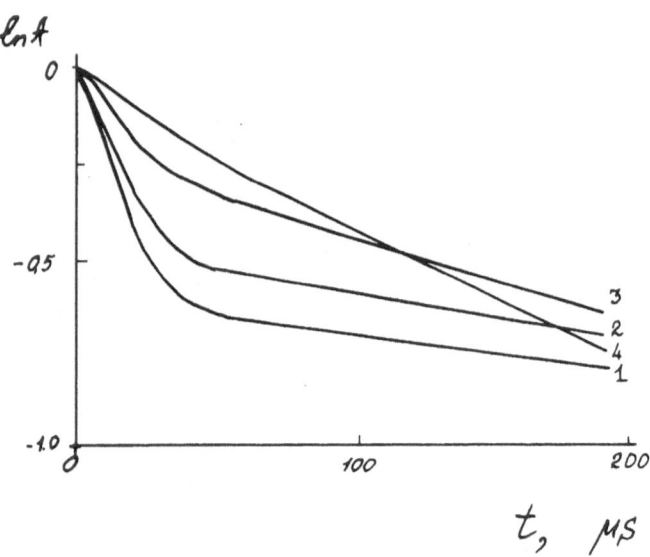

Fig. 1 FID of the MMA–MAA copolymer – FP plasticizer system with $b = 3.6$ measured on ^1H at 328 K. The heating time (min): 0(1), 5(2), 15(3), 120(4)

decreases. Hence the rate of the plasticizer diffusion and the polymer devitrification (the first stage of the process), which give rise to $p(2)$ and $T_2(2)$ increase, is higher than that of the gel formation (the second stage), which results in decrease of polymer molecules mobility. At $b = 3.6$ the mobility of macromolecules of the plastified fraction of the polymer, which is characterized by $T_2(2)$, decreases continuously (Fig. 2, curve 1). This indicates the closeness of the polymer devitrification and gel formation rates under these conditions.

Kinetics of the plastigel formation determined from changes in FID form measured on ^{19}F complement the process mechanism. In the studied systems only plasticizer molecules have ^{19}F nuclei. This allows the study of the mobility of the plasticizer molecules. ^{19}F FIDs can be approximated by the sum of two exponential functions. The long FID component corresponds to the "free" plasticizer, the molecules of which are located among the molecules of the nonvitrified polymer or among the fragments of the polymer and their aggregates. The short FID component corresponds to the "adsorbed" plasticizer whose molecules are in the vicinity of the surface of the vitrified polymer.

As indicated in Fig. 3 (curve 3), the contribution of the "adsorbed" plasticizer molecules $p(1)$ in FID measured 10–20 min from the beginning of the system heating increases drastically compared to $p(1)$ of the initial sample.

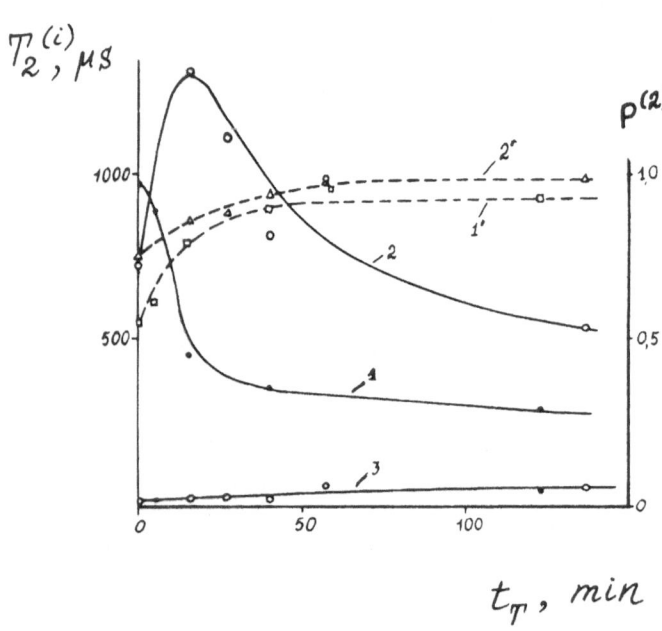

Fig. 2 $T_2(i)$ and $p(i)$ measured on ^1H of the MMA–AA–FP system vs heat duration at 328 K: $T_2(2)$ – (1, 2), $T_2(1)$ – (3), $p(2)$ – (1′, 2′), $b = 3.6$ – (1, 1′, 3), $b = 8$– (2, 2′, 3)

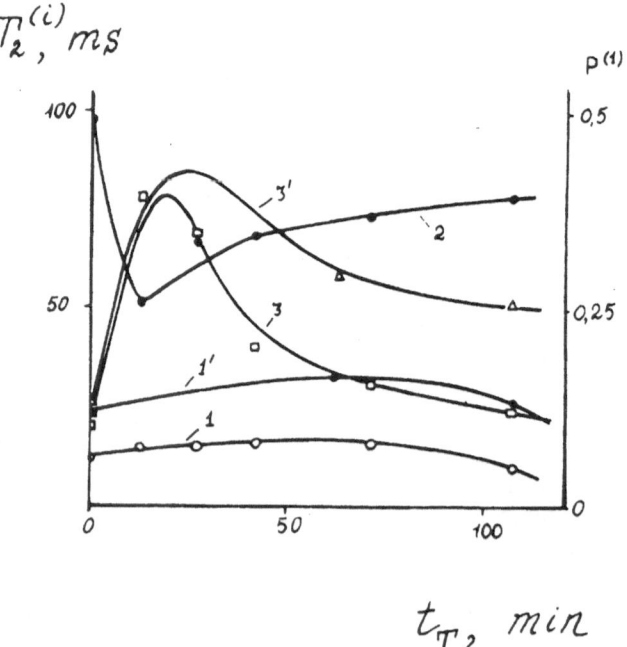

Fig. 3 $T_2(i)$ and $p(i)$ measured on ^{19}F of the MMA–MAA–FP system vs heat duration at 328 K: $T_2(1)$ – (1, 1′), $T_2(2)$ – (2), $p(1)$ – (3, 3′), $b = 3.6$ – (1, 2, 3), $b = 8$ – (1′, 3′)

Progr Colloid Polym Sci (1996) 102:47–50
© Steinkopff Verlag 1996

This is likely because of the augmentation of the polymer surface due to decomposition of the aggregates of latex particles to the primary ones.

The aggregates' decomposition likely starts just with the plastisol heating. Due to this decomposition the plasticizer penetrates vigorously into the surface layers of the polymer resulting in its devitrification. Referring to Fig. 3 (curve 3), in this process the fraction $p(1)$ of the adsorbed plasticizer molecules decreases and their mobility (according to $T_2(1)$) increases continuously (curve 1). In the end of the gel formation the mobility of the "adsorbed" molecules slightly decreases concurrently with the increase in $T_2(2)$. This is likely connected with the beginning of the third stage of the sol–gel transition: the rearrangement of the primary gel structure.

To accomplish the rearrangement of the primary gel structure, the samples were kept for 2 days at 293 K. As is evident from Table 1, in which the results of ^{1}H FID proceeding are listed, some polymer molecules in plastigel are vitrified or crystallized ("frozen" regions). The other FID component, which is characterized by the relaxation time of 100 μs, likely corresponds to the macromolecules fragments adjacent to the "frozen" regions. The parameters $p(3)$ and $T_2(3)$ of the third FID component characterize the "mobile" regions of the sample such as a highly elastic polymer and plasticizer.

We established that in the end of the second stage at $b = 3.6$ the polymer is practically homogeneous and highly elastic. Hence the inhomogeneity emerging at the third stage may be caused by significant rearrangement of the primary gel structure through the polymer crystallization in the presence of the plasticizer (then the microcrystals formed in this process play the role of nodes in a macromolecular network) and (or) through unfinished amorphous stratification of the polymer solvent at cooling resulting in formation of the vitrified polymer regions.

FIDs of these systems measured at 293 K on ^{1}H by the Carr–Purcell technique can be fitted by the sum of three exponential curves.

The $\lg(T_2(i))$ vs copolymer concentration dependencies (Fig. 4) can be fitted by fractured lines with drastic change in the slope of the lines at b of about 3.6. This indicates the change of the plasticizer condition at this point.

^{19}F FIDs of the samples with various plasticizer content are shown in Fig. 5. It is evident that FIDs of the

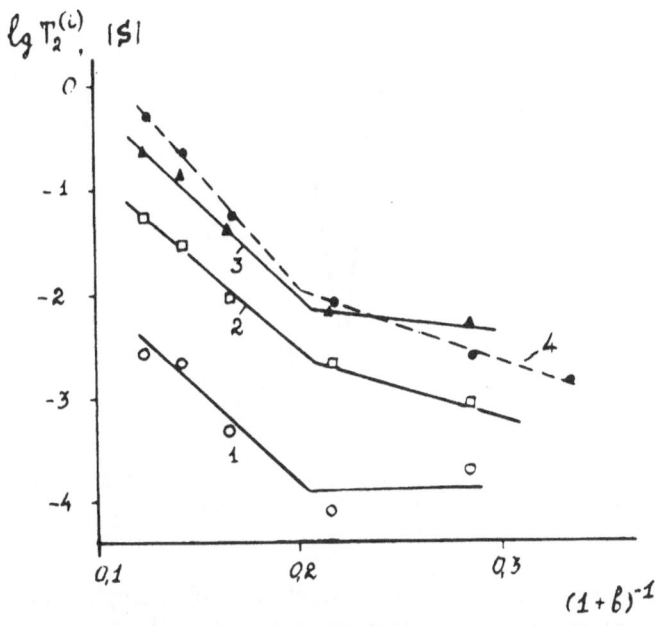

Fig. 4 $T_2(i)$ measured vs relative weight copolymer concentration $(1 + b)^{-1}$ in the plastigel at 293 K: $T_2(1) - (1)$, $T_2(2) - (2)$, $T_2(3) - (3, 4)$, FID (^{1}H) $- (1, 2, 3)$, FID (^{19}F) $- (4)$

Table 1 Relaxation times $T_2(i)$ and fractions $p(i)$ which characterize short ($i = 1$), middle ($i = 2$), and long ($i = 3$) ^{1}H FID components at 293 K vs polymer concentration $(1 + b)^{-1}$ in the sample

$(1+b)^{-1}$	$-\lg T_2^{(1)}$ [s]	$-\lg T_2^{(2)}$ [s]	$-\lg T_2^{(3)}$ [s]	$p^{(1)}$	$p^{(2)}$	$p^{(3)}$
0			0.548			1.0
0.125	2.590	1.290	0.655	0.14	0.48	0.38
0.143	2.685	1.555	0.905	0.13	0.64	0.23
0.167	3.340	2.071	1.425	0.13	0.55	0.31
0.217	4.121	2.697	2.182	0.11	0.52	0.37
0.285	3.721	3.092	2.333	0.21	0.76	0.03

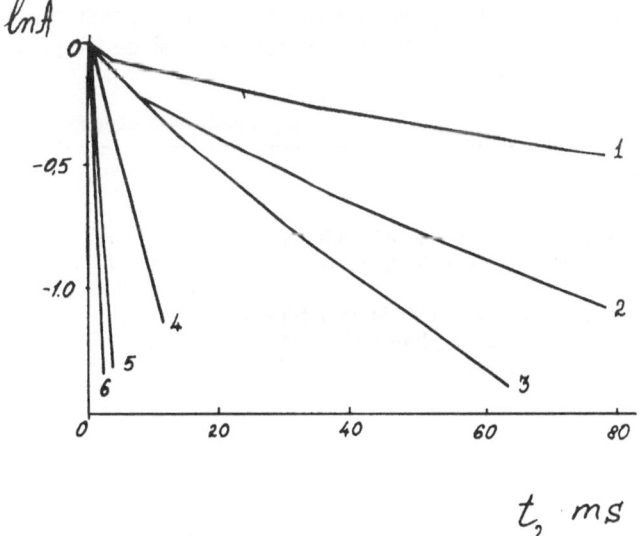

Fig. 5 ^{19}F FID of the MMA–MAA–FP system at 293 K at the following plasticizer – polymer ratios b: 10(1), 8(2), 4(3), 3.6(4), 2(5), and 1(6)

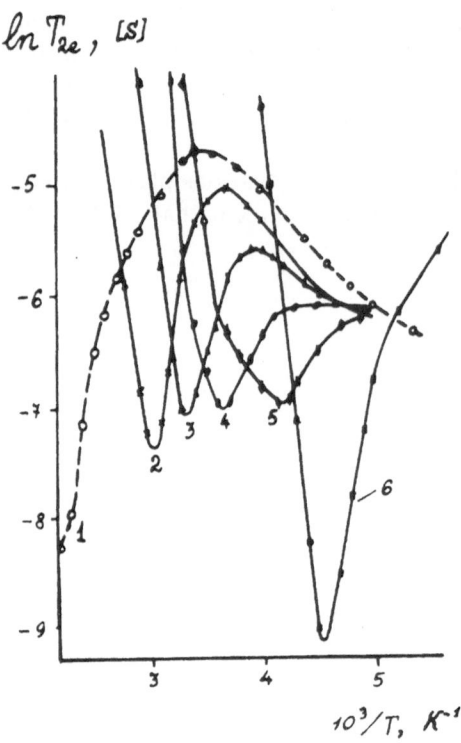

Fig. 6 Temperature dependencies of $\lg(T_{2e})$ of the MMA-MAA - FP system at b: 0(1), 1(2), 2(3), 4(4), 8(5), and ∞(6)

samples with $b \leq 3.6$ are characterized by a single relaxation time indicating the fact that at $b \leq 3.6$ the plasticizer forms a uniphase whereas at $b > 3.6$ it forms a multiphase system.

Complementary information about the relationship between molecular dynamics and plastigel structure was obtained by multipulse NMR techniques which allow measurements of molecular motion in a wide range of correlation times [6].

In the temperature range of (180–440) K a low temperature branch with $E_a = 8$ kJ/mol, which corresponded to motion of the lateral groups of the polymer chain (CH_3-) and a high temperature one, which corresponded to the small-scale segmental motion with $E_a = 42$ kJ/mol consistent with E_a of hydrogen bonds of carboxylate groups of MAA chains, were measured on the copolymer.

From slopes of high temperature branches of $\lg(T_{2e})$ vs T^{-1} determined from signals on ^{1}H and ^{19}F nuclei of the plasticizer, $E_a = 83$ kJ/mol and $\tau = 10^{-18}$ s which characterized the motion of FP molecules were calculated.

The $\lg(T_{2e})$ vs T^{-1} curves shifted continuously to the low temperature range toward the relevant curve for the plasticizer (Fig. 6).

Conclusion

It is shown that sol–gel transition in the systems based on the methyl methacrylate–methacrylic acid copolymer proceeds via three stages. At the first stage diffusion of the swelling agent caused devitrification of the copolymer. Concurrently, the aggregates of latex particles disintegrated on the macroscopic scale. On the second stage gel with non-equilibrium structure was formed. Then the structure was rearranged during the third stage. The relationship between the rates of the processes which control these three stages depends on the extender concentration. When the sol–gel transition has been completed, the system was polyphase. It was constituted of highly elastic, crystalline (or vitreous), and transition phases. The ultimate degree of compatibility of a polymer-swelling agent pair can be calculated by plotting logarithm of the spin–spin relaxation time T_2 vs concentration.

References

1. Yanovsky UG, Vasin AV, Vinogradov GV et al (1982) Vysokomol Soedin 24:2563
2. Erofeev LN, Vetrov OD, Shumm BA et al (1977) Pribory i tekhnika eksperimenta (Russia) 2:145–149
3. Carr HJ, Purcell EM (1954) Phys Rev 94:630
4. Mansfield P, Ware P (1968) Phys Lett 23:421
5. Rhim W-K, Pines A, Waugh JS (1971) Phys Rev 3:684
6. Manelis GB, Erofeev LN, Provotorov BN et al (1989) Sov Sci Rev B Chem 14:1–92

Progr Colloid Polym Sci (1996) 102:51–56
© Steinkopff Verlag 1996

B. Ginzberg
S.A. Bilmes

Titania sols and gels synthesized from reverse micelles

Dr. B. Ginzberg (✉)
Departamento de Química
Facultad de Ingeniería
Universidad de Buenos Aires
Paseo Colón 850
1043 Buenos Aires, Argentina

S.A. Bilmes
INQUIMAE
Departamento de Química Inorgánica
Analítica y Química-Física
Facultad de Ciencias Exactas y Naturales
Universidad de Buenos Aires
Ciudad Universitaria Pab II
1428 Buenos Aires, Argentina

Abstract TiO_2 nanoparticles are synthesized in water/AOT/isooctane reverse micelles by hydrolysis of titanium (IV) n-butoxide. The molar ratios of the three components defines the size of nanoparticles as determined from the uv-visible absorption onset of the disperse system. Depending on the relative content of each component this sol undergoes a gel phase. The kinetics of sol–gel process is monitored for different $Ti(BuO)_4$/water/AOT molar ratios, and as a function of temperature by static viscosity, and by the optical density of the system in the spectral range where light scattering is the dominant optical process. The water content of the system and the $Ti(BuO)_4$/water molar ratio have great influence on the kinetics of the gelation process. An activation energy of (91 ± 9) kJmol^{-1} was obtained from Arrhenius plots.

Key words Titanium dioxide – reverse micelles –nanoparticles – gel – sol–gel

Introduction

Titanium dioxide is an interesting non-toxic semiconductor material due to its high stability in aqueous solutions. It is widely employed in paints [1], antireflective coatings [2], photocatalyst for solar energy conversion [3,4] and for water detoxification [5,6]. Nanometer-sized semiconductor clusters with diameters between 1 and 10 nm fall into the transition state between molecular and bulk material properties, and exhibit quantization effects [7] which lead to enhanced catalytic behavior [8,9] and non-linear optical properties [8–10]. These Q-size particles are characterized by a blue shift of the absorption edge due to the increasing energy of the band gap as the particle radius decreases [8, 9, 11].

Several routes have been employed for the synthesis of titanium dioxide nanoparticles from liquid solutions of titanium precursors [12–14]; the main problem is the difficulty to prevent the particle growth. Water-in-oil dispersions allow to confine the reaction in the micelle cavities, hindering the growing process. The formation of semiconductor clusters in the system is controlled by the reactant distribution over the micelles and by the dynamics of intermicellar exchange. Several Q-size semiconductors have been synthesized by this method [15–17].

In order to employ these particles as catalysts or coatings, it is necessary to immobilize them onto an appropriate substrate. One way to accomplish this task is by obtaining a gel from the sol, which is further attached to the substrate by spin coating or dipping [18, 19]. The sol–gel process has been widely studied in connection to the synthesis of SiO_2 glasses and TiO_2 films and membranes [20, 21].

In this work, the synthesis of TiO_2 nanoparticles from $Ti(BuO)_4$ in water/AOT/isooctane reverse micelles and the sol–gel transition for this system are analyzed as a function of the relative content of each component, and as a function of temperature in order to determine the optimal conditions for the preparation of gel-precursors for coatings on glasses and metals. The molar ratios of the three components determines the size of nanoparticles and the possibility of further evolution to a gel phase. By increasing the water content in the disperse system, the radius of Q-size TiO_2 grows and the kinetics of the sol–gel transition becomes faster.

Materials and methods

Reverse micelles were prepared by adding water (distilled and further purified by a Milli-Q system) to AOT (sodium bis(2-ethylhexyl) sulfosuccinate; Sigma, 99%) dissolved in isooctane (Fluka, HPLC grade). In all cases the [water]/[AOT] molar ratio, w, was lower than 2.4. Titania sols were synthesized by adding titanium butoxide (Strem Chemicals) to the micellar system under stirring. By this procedure, clear sols are obtained. Different molar ratios of [Ti(BuO)$_4$]/[water]/[AOT] were employed. Depending on the molar ratio of the components some of these sols develop a gel phase.

The size of semiconductor nanoparticles in the sol phase was estimated using the model of Brus [11] from the absorption edge in the uv-visible absorption spectra. These were taken with a diode array absorption spectrophotometer (Hewlett–Packard, 8452A).

The sol–gel transition kinetics was monitored in a thermostated cell following the time evolution of the static viscosity measured with a rotational viscometer (Brookfield LVDVII) with digital data acquisition. Measurement of the optical density at 800 nm as a function of time was used as an alternative method which allows to work with a shorter time window.

Results and discussion

Titania sols

Titanium butoxide is highly soluble in hydrocarbons. When introduced in the w/o microemulsion it dissolves in the continuous phase. By interaction with water enclosed in the surfactant cavity it undergoes a partial or a complete hydroxylation:

$Ti[O(CH_2)_3CH_3]_4 + xH_2O$

$\quad = Ti[(OH)_x(O(CH_2)_3CH_3)_{(4-x)}]$

$\quad\quad + xCH_3(CH_2)_2CH_3OH$ (1)

$Ti[(OH)_x(O(CH_2)_3CH_3)_{(4-x)}] + (4-x)H_2O$

$\quad = TiO_2 + (4-x)CH_3(CH_2)_2CH_3OH + 2H_2O$ (2)

Reaction (1) involves a nucleophilic attack of the Ti atom by the oxygen atom of a water molecule, the transfer

of a proton from the water to an $-O(CH_2)_3CH_3$ group of the metal, and the release of the resulting alcohol molecule.

The hydrolysis is spatially restricted to the surface of the microcavities, and therefore the size and shape of the semiconductor particle is determined by the size and shape of the reverse micelle. Micelle size depends linearly on the water-to-surfactant molar ratio, w [22]. The water pool radius, R_w, derived from the micelle Stokes radius after deducting the size of the surfactant chain, increases with water contents [23]; for not too low w values the relationship is linear:

$$R_w(\text{Å}) = 1.5w .$$ (3)

For AOT–water–isooctane the relation (3) has been experimentally proven by photon correlation spectroscopy, fluorescence anisotropy, SAXS and SANS [22–24]. However, all our experiments were performed at very low w values; it has been shown that in the range $w < 9$, R_w increases slowly with increasing w [22–24]. A reasonable assumption for our conditions is $R_w = 15 \pm 4$ Å. In this case, the aggregates behave like rigid macromolecules with water highly structured in the aggregates by hydrogen bonds stabilized by the strong dipole moments of the head groups of AOT. It is thus concluded that R_w changes should not greatly influence our results; on the other hand, water availability, as measured by w is a crucial parameter for the hydrolysis (1). Therefore, we shall center all the following discussion on the influence of w on the process under study.

For semiconductor particles, quantum mechanical calculations [11, 25, 26] and experimental observations [12, 27, 28] conclude that the energy level of the first excited state of the exciton increases with decreasing particle size, leading to a blue-shift in the absorption spectrum. Particle growth results in spectral shifts from an absorption onset in the UV for dimers and oligomers to that corresponding to the bulk material. For a sphere of radius R and dielectric constant ε, Brus [11] calculates the energy of the lower excited state, E^*, in a cluster from a Hamiltonian which takes into account both Coulomb interaction between hole and electron, and the quantum localization energy of a "pseudoelectron" of effective mass m_e^*:

$$E^* \cong E_g + (h^2/8R^2)[(1/m_e^*) + (1/m_h^*)] - (1.8e^2/\varepsilon R) .$$ (4)

The Coulomb term gives rise to shifts of E^* to lower energy that are proportional to R^{-1}, while the shift of E^* to higher energy due to quantum localization varies as R^{-2}. This latter term dominates the overall particle size effect and, for small values of R, the apparent band gap is larger than that of the bulk material, E_g. Equation (4) may be used to estimate the radius of nanoparticles, from the absorption onset in the UV-visible absorption spectra.

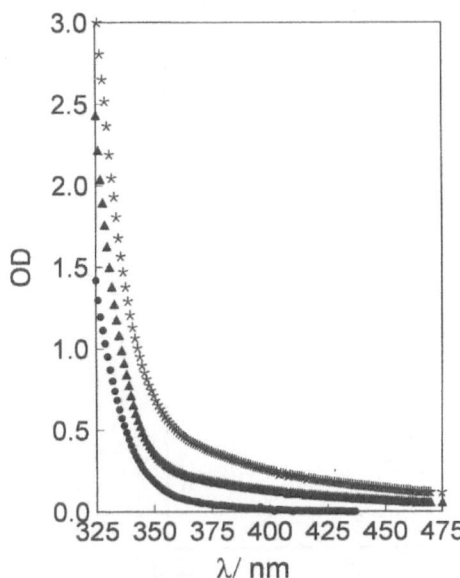

Fig. 1 Absorption spectra of titania sols in reverse micelles of water in isooctane. (∗) [Ti(BuO)$_4$] = 0.01 M, [AOT] = 0.1 M and [H$_2$O] = 0.02 M; (▲) [Ti(BuO)$_4$] = 0.01 M, [AOT] = 0.1 M and [H$_2$O] = 0.01 M; (●) [Ti(BuO)$_4$] = 0.006 M, [AOT] = 0.4 M and [H$_2$O] = 0.012 M

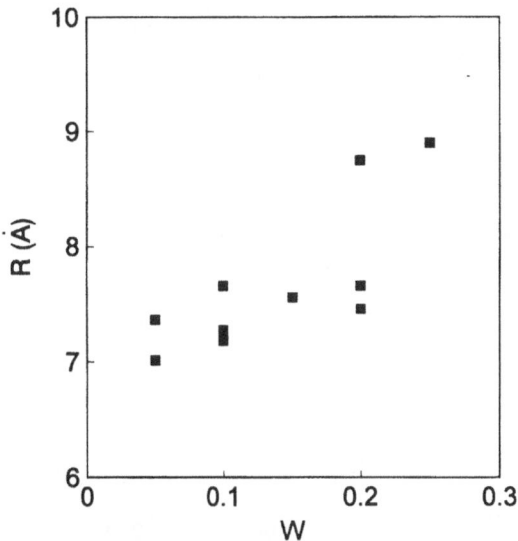

Fig. 2 Dependence of the TiO$_2$ nanoparticle radius, R, on the parameter w. Data were obtained for different compositions of the disperse system

Figure 1 shows the uv-visible absorption spectra of titania sols synthesized in reverse micelles with w between 0.05 and 0.25. Although butanol, AOT and isooctane have no significant absorption in the 300–400 nm range, in these experiments the reference spectrum is that of a system composed of water/AOT/isooctane and butanol in the same concentrations as the TiO$_2$ containing sample. The onset of absorption (determined by the linear extrapolation of the steep part of the uv absorption toward the base line [28]) is blue shifted as w decreases due to the low availability of water, which limits the growth of semiconductor particles.

Even though we do not have information about the composition and structure of the particles formed by reaction Ti(OBu)$_4$ and H$_2$O, it is reasonable to assume that the particles may be considered as "TiO$_2$ nanocrystals", irrespective of whether they contain OH, n-Butoxide and/or AOT bound to the Ti(IV). The similarity of the spectral changes attending the reaction in the micelles and those observed in solution supports this description. Therefore, we have estimated the radius R of the particles from the absorption onset using Eq. (4) with the following parameters: $[(1/m_e^*) + (1/m_h^*)]^{-1} = 1.63\ m_e$ [12, 13] ($m_e = 9.110^{-28}$ g is the electron rest mass); $\varepsilon = 184$ [29], and $E_g = 3.2$ eV for anatase [30]. From Fig. 2 it can be seen that the TiO$_2$ particles synthesized in reverse micelles have calculated radii between 7 and 9 Å. These values

represent a rough approximation, taking into account the ambiguity in the determination of the absorption onset, and the 50% spread of values for the effective masses reported in the literature [13].

Sol–gel transition

Titania sols may be stable for several months or may evolve to a gel state depending on the Ti(BuO)$_4$/water/AOT molar ratio. The time evolution of the sols was monitored by measuring the static viscosity, η, as a function of time for different compositions of the disperse system. Figure 3 shows the static viscosity – time curves for different Ti(BuO)$_4$/water/AOT molar ratios. In each plot the concentration of one component is varied while the others either remain constant or are adjusted in order to maintain a constant [water]/[Ti(BuO)$_4$] molar ratio equal to 2. The latter corresponds to the stochiometric molar ratio for the complete hydrolysis to TiO$_2$ according to the reaction scheme represented by Eqs. (1)–(2). For shorter times, the viscosity remains constant and with a low value until a sudden increase is noticed. This abrupt change in the viscosity is related with the growth of clusters by aggregation of particles, and/or condensation reactions. Links are formed between clusters upon collision of them, and the gel appears when the last link between two large clusters is formed, leading to a continuous solid network. The gel point, t_g, can be defined as corresponding to a certain value of viscosity, and although it is not an

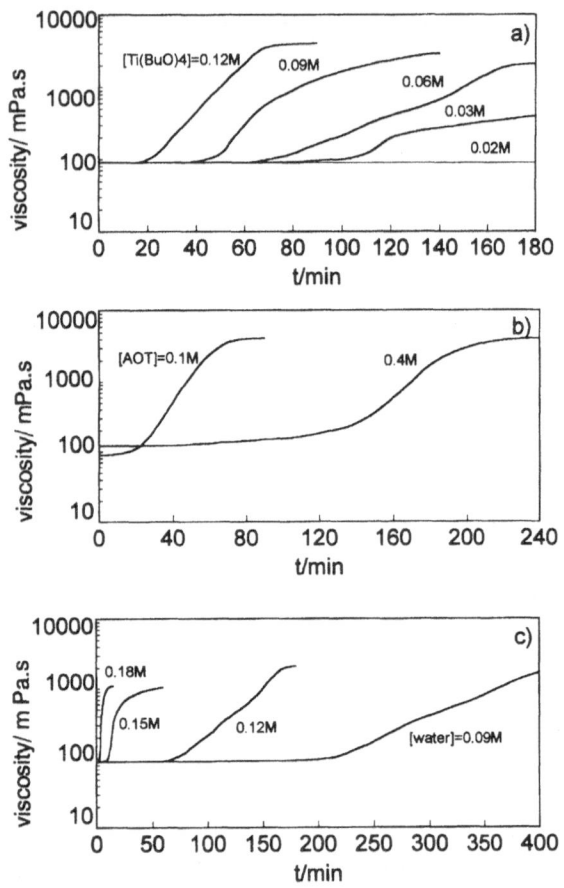

Fig. 3 Static viscosity, η, vs. time plots for different compositions of the disperse system: a) $[AOT] = 0.1$ M; $[H_2O]/[Ti(BuO)_4] = 2$; for the $[Ti(BuO)_4]$ values indicated on each curve. b) $[Ti(BuO)_4] = 0.12$ M; $[H_2O] = 0.24$ M; for the $[AOT]$ values indicated on each curve. c) $[AOT] = 0.1$ M; $[H_2O]/[Ti(BuO)_4] = 2$; for the $[H_2O]$ values indicated on each curve

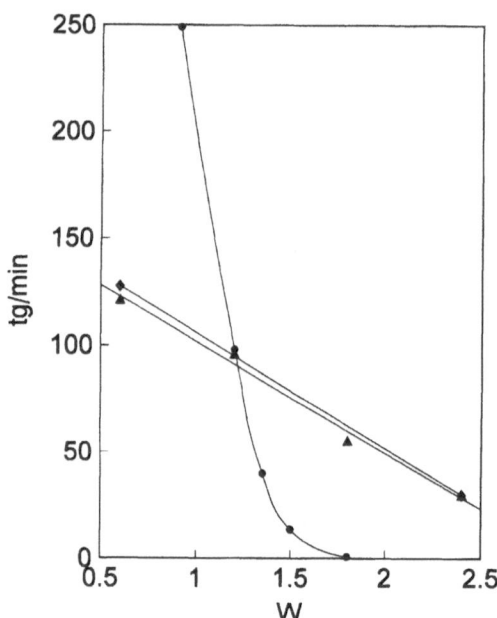

Fig. 4 Dependence of the gelation point (defined as the time elapsed to reach $\eta = 200$ mPa.s) on the parameter w, for the variables corresponding to Fig. 3. i) (●) variable is $[AOT]$, ii) (▲) variable is $[H_2O]$; iii) (◆) variable is $[Ti(BuO)_4]$

unambiguous magnitude, it can be used to identify the gel point, t_g, in a crude way [20].

From Fig. 3a it can be seen that as $[Ti(BuO)_4]$ and w decrease, t_g increases. For alcohoxide concentrations lower than 0.02 M in 0.1 M AOT, the sol phase is stable at least for 6 months. The gel point also increases by increasing the surfactant concentration, i.e. by decreasing the water availability in each micelle, as shown in Fig. 3b. The importance of water availability is further demonstrated by Fig. 3c, which shows that the $[water]/[Ti(BuO)_4]$ molar ratio has a pronounced effect on t_g. Sols formed with $[water]/[Ti(BuO)_4] < 1$ did not form a gel phase.

Figure 4 compares the influence of w on t_g for three different cases. The gelation time, t_g, was defined as the time required to achieve a viscosity value $\eta = 200$ mPa.s. In (i) (full line) w is changed by varying $[AOT]$, at constant $[H_2O]$, whereas in (ii) (dashed line) $[H_2O]$ is changed at constant $[AOT]$. The $[Ti(BuO)_4]$ is maintained constant

in both cases. In case (i) gelation time decreases linearly with w, whereas changes in water concentration in the range 0.09–0.17 M produce a drastic effect, again illustrative of the limitation by water availability. Curve (iii) shows the influence of w in a set of experiments in which $[H_2O]$ and $[Ti(BuO)_4]$ were changed simultaneously with constant $[AOT]$. In this case water availability is not limited, and the slow linear relationship of curve (i) is obtained again.

The sol–gel transition can also be monitored by measuring the optical density of the samples as a function of time. For the sol phase, the spectrum is characterized by the absorption of semiconductor particles below 400 nm. Figure 5a shows the spectra taken at time intervals of 10 min for the system with molar ratio composition $[Ti(BuO)_4]:[H_2O]:[AOT] = 0.12:0.24:0.4$. For times lower than 100 min there is a shift of the TiO_2 absorption onset to lower energies due to the increase in the size of nanoparticles, probably by aggregation or by condensation reactions involving partially hydroxylated $Ti(OH)_x$ species. At the beginning of the process the particle size may be derived from absorption measurements; at longer times, however, scattering effects become very important, and further evolution of individual particle size may not be followed. Scattering is evident by the increase of optical density in the visible range according to the λ^{-4} law.

Progr Colloid Polym Sci (1996) 102:51−56
© Steinkopff Verlag 1996

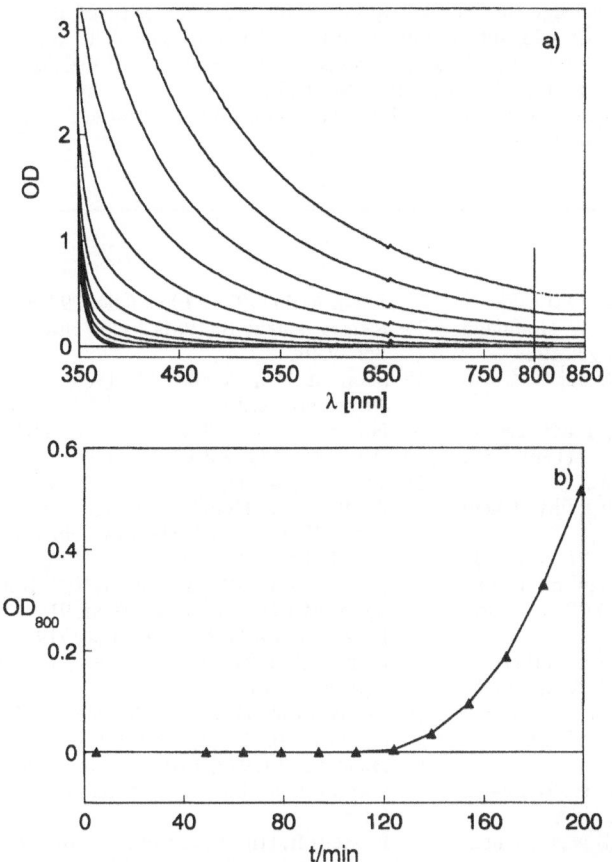

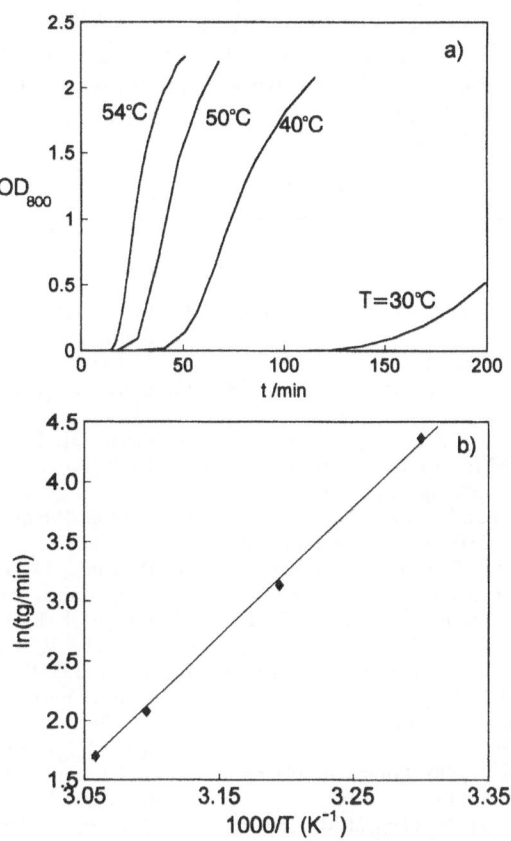

Fig. 5 a) Time evolution of the uv-visible spectra for the system with composition $[Ti(BuO)_4]:[H_2O]:[AOT] = 0.12$ M: 0.24 M: 0.4 M. Spectra were taken at fixed time intervals of 10 min. b) Plot of the optical density at 800 nm, OD_{800}, vs. time from the spectra depicted in (a)

Fig. 6 a) Optical density at 800 nm vs. time for the $[Ti(BuO)_4]:$ $[H_2O]:[AOT] = 0.12:0.24:0.4$ system at the indicated temperatures. b) Arrhenius plot for the system composition of (a)

A plot of the optical density at 800 nm, OD_{800}, is shown in Fig. 5b, where a sudden increase in the measured magnitude is noticed for $t > 130$ min. The coincidence of the sudden increases of both viscosity and optical density at 800 nm is also noticed for other compositions of the disperse system. This allows to assign the onset of OD_{800} increase to the gelation time t_g.

The kinetics of sol−gel transition is also dependent on temperature. Figure 6 shows the OD_{800} − time dependence for the $[Ti(BuO)_4]:[H_2O]:[AOT] = 0.12:0.24:0.4$ system at different temperatures. The Arrhenius plot (Fig. 6b) for this system gives an apparent activation energy of (91 ± 9) kJ·mol^{-1}, in good agreement with that found in the literature for silica gels [20, 31]. However, this apparent activation energy cannot be ascribed to any particular reaction, as gelation depends in a complicated way on the rates of hydrolysis, condensation, and diffusion of clusters.

The mechanism of gelation in the micellar system is not clear. It is tempting to conclude that butoxide groups bound to polymerized titanium ions may "solubilize" the nanoparticles in the continuous medium, thus allowing the encounter between hydrolyzed particles.

Conclusions

Sols and gels of titanium dioxide can be successfully synthesized in reverse micelles. The hydrolysis reaction of $Ti(BuO)_4$ with water structured in reverse micelles produces $Ti[(OH)_x(O(CH_2)_3CH_3)_{(4-x)}]$ which behaves as "TiO_2 nanoparticles". The radii of the nanocrystals are determined by the size of the reverse micelle where the hydrolysis takes place. For systems with [water]/$[Ti(BuO)_4] \geq 1$ aggregation and/or condensation reactions take place, leading to a gel phase. The kinetics of sol−gel transition strongly depends on the water content of the system. Gelation probably proceeds through $Ti[(OH)_x(O(CH_2)_3CH_3)_{(4-x)}]$ species with $x \leq 2$ which should be soluble in the hydrocarbon. Arrhenius plot for the sol–gel transition gives an apparent activation energy of $(91 \pm 9)\,kJ \cdot mol^{-1}$.

Acknowledgments Work supported by the University of Buenos Aires (Ex 022; In023), and by Fundación Antorchas. The authors thanks to Luis Baikauskas for his assistance in part of the experiments, and to Miguel Blesa for fruitful discussions and critical reading of the manuscript. B.G. thanks David Kurlat for encouragement and support.

References

1. Judin VP (1993) Chemistry in Britain, p 503
2. Samuneva B, Kozhukarov V, Trapalis Ch, Kranols R (1993) J Mat Sci 28:2353
3. Honda K (1972) Nature 238:37
4. Memming R (1991) In: Pelizzetti E, Schiavello M (eds) Photochemical Conversion and Storage of Solar Energy. Kluwer (Holland) p 193
5. Bahnemann DW, Cunningham J, Fox MA, Pelizzetti E, Pichat P, Serpone N (1993) In: Helz GR, Zepp RG, Crosby DG (eds) Aquatic and Surface Photochemistry; Lewis Publ, p 261
6. Leguini O, Oliveiros E, Braun A (1993) Chem Rev 93:671
7. Papavassiliou G (1981) Solid State Chem 40:330
8. Henglein A (1988) Topics in Current Chemistry 143:113
9. Kamat P (1991) In: Grätzel M, Kalyansuaram K (eds) Kinetics and Catalysis in Microheterogeneous Systems; Marcel Dekker, USA, p 375
10. Chemela DS, Müller DA (1985) J Opt Soc Am B2:1155
11. Brus L (1986) J Phys Chem 90:2555
12. Kormann C, Bahnemann DW, Hoffmann MR (1988) J Phys Chem 92:5196
13. Bahnemann D (1993) Israel J Chem 33:115
14. Micic OI, Rajh T, Comor MI, Zec S, Nedeljkovic JM, Patel RC (1991) Mat Res Soc Symp Proc 206:127
15. Petit C, Pileni MP (1989) J Phys Chem 92:2282
16. Papoutsi D, Lianos P, Yianoulis P, Koutsoukos P (1994) Langmuir 10:1684
17. Joselevich E, Willner Y (1994) J Phys Chem 98:7628
18. Smestad G, Bignozzi C, Argazzi R (1994) Solar Energy Mats and Solar Cells, 32:259
19. Kim DH, Anderson MA (1994) Env Sci and Technol 28:479
20. Brinker CJ, Scherrer GW (1989) Sol–gel Science; Academic Press, USA
21. Sanchez C, Ribot F (1994) New J Chem 18:1007
22. Zulauf M, Eike HF (1979) J Phys Chem 83:480
23. Pileni MP (1993) J Phys Chem 97:6961
24. Keh E, Valeur B (1981) J Coll Interface Sci 79:465
25. Schmidt HM, Weller H (1986) Chem Phys Lett 129:615
26. Hagfelt A, Lunell S, Siegbahn HOG (1994) Int J Quant Chem 49:97
27. Weller H, Schmidt HM, Koch U, Fojtik A, Baral S, Henglein A, Kunath W, Weiss K, Dieman E (1986) Chem Phys Lett 124:557
28. Bahnemann DW, Kormann C, Hoffmann MR (1987) J Phys Chem 91:3789
29. Parker A (1961) Phys Rev 124:1719
30. Bickley RI (1978) Chem Phys of Solids and their Surfaces 7:118
31. Colby MW, Osaka A, Mackenzie JD (1988) J Non-Cryst Solids 99:129
32. Hiemenz P (1977) Principles of Colloid and Surface Chemistry, Marcel Dekker, USA
33. Rossetti R, Hull R, Gibson JM, Brus LE (1985) J Chem Phys 83:1406

Progr Colloid Polym Sci (1996) 102:57–63
© Steinkopff Verlag 1996

Gels with magnetic properties

L. Barsi
A. Büki
D. Szabó
M. Zrinyi

L. Barsi · A. Büki · D. Szabó
Prof. Dr. M. Zrínyi (✉)
Department of Physical Chemistry
Technical University of Budapest
1521 Budapest, Hungary

Abstract Materials producing strain in magnetic field are known as magnetoelastic or magneto strictive materials. A new type has been developed by preparing magnetic field sensitive gels, called ferrogels. Single domain, magnetic particles of colloidal size are incorporated into chemically cross-linked polyvinyl-alcohol hydrogels. The finely distributed colloidal particles having superparamagnetic behavior couple the shape of the gel to the non-uniform external magnetic field. We have shown that ferrogels undergo quick and reversible shape transformation by changes in external magnetic field. Elongation, contraction and bending can be realized by proper arrangement of external magnetic field. Unidirectional deformation measurements have been performed and an equation for the uniaxial magnetoelastic properties has been derived and compared with experimental results.

Key words Magnetic gels – superparamagnetic behavior – filled networks – magneto-elasticity – magnetostriction

Introduction

In the last decades many research groups have been studying the elastic and swelling behavior of hydrogels for possible technological and biomedical applications. Special attention has been devoted to stimuli-responsive gels which may swell or deswell with small changes of environmental conditions [1, 2]. Volume phase transition in response to infinitesimal change of external stimuli like pH, temperature, solvent composition, electric field and light has been observed in various gels [1–3]. Their application in devices such as actuators, controlled delivery systems, sensors separators and artificial muscles are in progress. Attempts at developing stimuli-responsive gels for technological purposes are often complicated by the fact that volume phase transition is kinetically restricted by both the collective diffusion of the chains and the friction between polymer network and swelling agent. This disadvantage often hinders the effort of designing optimal gels for different applications.

In order to accelerate the response of an adaptive gel to stimuli, magnetic field sensitive gels called ferrogels have been developed [4, 5]. For a ferrogel not the volume, but its shape changes in non-uniform magnetic field. Shape distortion occurs instantaneously and disappears abruptly when the external magnetic field is removed.

This paper is concerned with the magneto-elastic properties of ferrogels in order to be effectively controlled and understood well enough to develop analytical models.

Experimental part

Preparation of ferrogels

A magnetic field sensitive gel is a special type of filler loaded gel, where the finely divided filler particles have strong magnetic properties. Preparation of a ferrogel does not require a special polymer or a special type of magnetic particles. As a polymer network one may use every flexible chain molecule which can be cross-linked. The filler particles can be obtained from ferromagnetic materials.

Preparation of a ferrogel is similar to that of other elastomeric networks. One can precipitate well-dispersed particles in the polymeric material. The "in situ" precipitation can be made before during and after the cross-linking reaction [6]. According to another method the preparation and characterization of colloidal magnetic particles are made separately, and the cross-linking takes place after mixing the polymer solution and the magnetic sol together [7]. In this paper chemically cross-linked polyvinyl alcohol hydrogel filled with magnetite particles is reported. At first, magnetite (Fe_3O_4) sol was prepared from $FeCl_2$ and $FeCl_3$ in aqueous solution. In order to counter-balance the van der Waals attraction and the attractive part of magnetic dipole interactions, colloidal stability has been maintained by a small amount of $HClO_4$ which induced peptization. Then the stabilized magnetite sol having a concentration of ~ 10 m% was mixed with polyvinyl alcohol solution. Polyvinyl alcohol (PVA) is a neutral water-swollen polymer which reacts under certain conditions with glutardialdehyde (GDA) resulting in chemical cross-linkages between PVA chains. The cross-linking density can be conveniently varied by the amount of GDA relative to the vinyl alcohol [VA] units of PVA chains. We have prepared weak PVA gels in order to allow the effect of magnetic interaction on deformation to develop as far as possible. The ratio of [VA] units to [GDA] molecules was varied between $100 \leq$ [VA]/[GDA] ≤ 400. A more detailed description of chemical procedure can be found in our previous paper [5].

Ferrogels of defined shape have been prepared. Cylindrical samples having a diameter of 1–2 cm and length of 10–20 cm were obtained. These gel tubes were used for the magnetoelastic investigations.

The ferrogels are characterized by the following symbols: sample name/polymer concentration (given in wt.%) at which cross-links were introduced/the ratio of vinyl alcohol units to the cross linking molecules/wt.% of filler particles in the gel. For example, a gel FG/6.3/300/4.25 means a ferrogel (FG) which was prepared at a polymer solution of 6.3 wt% of polyvinyl alcohol, the ratio of vinyl alcohol monomer units to the cross-linked molecules is 300, the magnetite content at preparation is 4.25 wt.%.

Unidirectional magneto-elastic measurements

In order to study the elastic response of ferrogels to magnetic field, cylindrical gel tubes were suspended in water

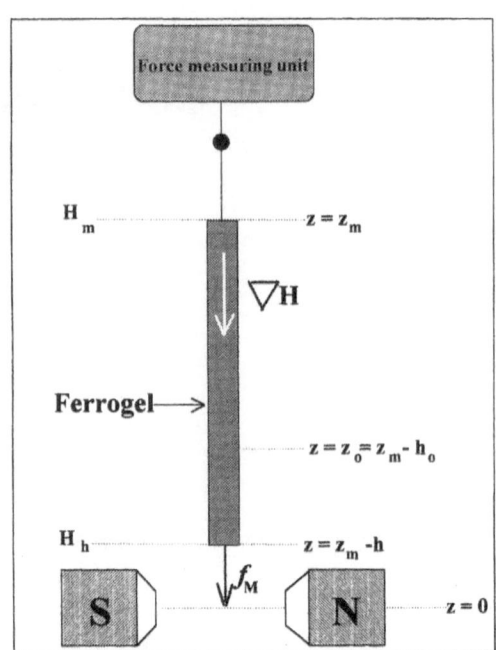

Fig. 1 Schematic diagram of experimental set up to study the magneto-elastic properties of ferrogels

vertically between plan-parallel poles of an electromagnet. The position of the top surface of gels was fixed by a rigid copper thread at the position designated by z_m on Fig. 1. The copper thread was connected to a force measuring unit in order to establish the force developed in the ferrogel due to interaction with magnetic field. The lower end of the suspended gel in the absence of external magnetic field is characterized by the position, z_0 measured from the $z = 0$ line which corresponds the axis at the face-to-face plan parallel magnetic poles of electromagnet. The magnetic field was induced by a steady current flowing through the electromagnets. The intensity of the current was varied between 0 and 10 A by an electronic power supply (FOK-GYEM TR-9177, Hungary) and the voltage was kept constant 40 V. The magnetic field vector, $\vec{B}(z)$ (sometimes called the magnetic induction or magnetic flux density) varies along the gel axis. This dependence was measured by a Teslameter (Phywe System Gmbh, Germany) at different steady current intensities.

The highest field strength between the poles of electromagnets we used was 840 mT. Figure 2 shows the dependence of magnetic field strength as a function of z, measured at different steady current intensities.

In order to determine the elongation-stress dependence experiments were monitored by a digital video system. A CCD camera with a 1/3-inch video chip has been connected to a PC through a real time video digitizer card. The movement of the end of gel cylinder was followed on the magnified picture by a homemade program, which has

Progr Colloid Polym Sci (1996) 102:57–63
© Steinkopff Verlag 1996

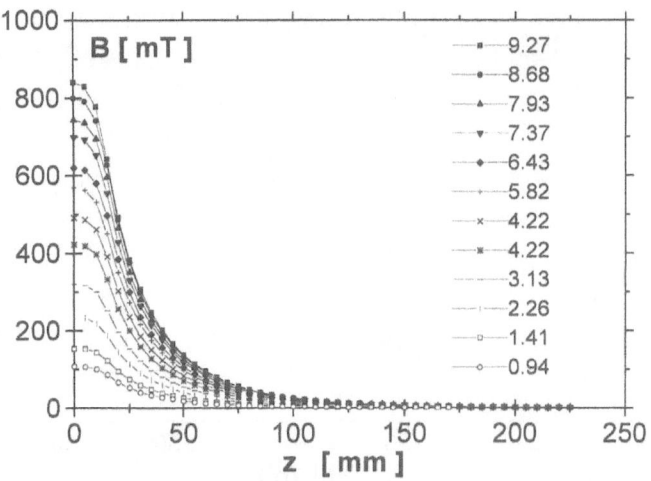

Fig. 2 Dependence of magnetic induction on the distance measured from the axis at face-to-face plan-parallel magnetic poles. The steady current intensities are indicated in the figure. Lines are a guide for the eye

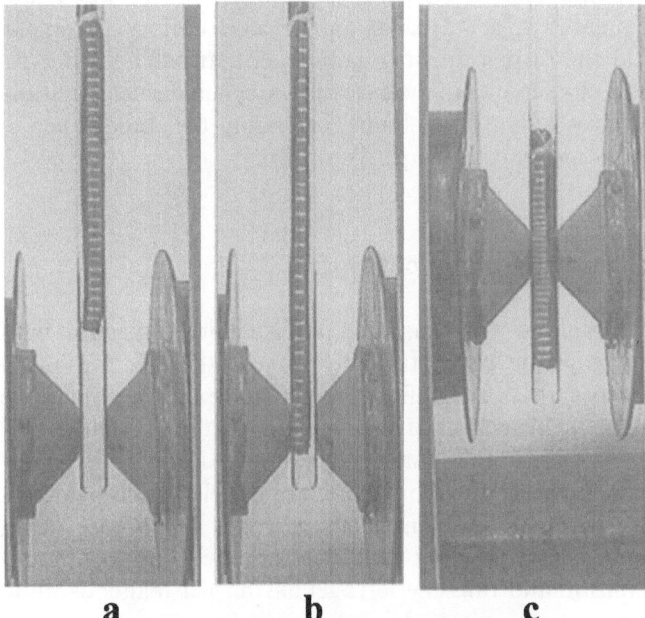

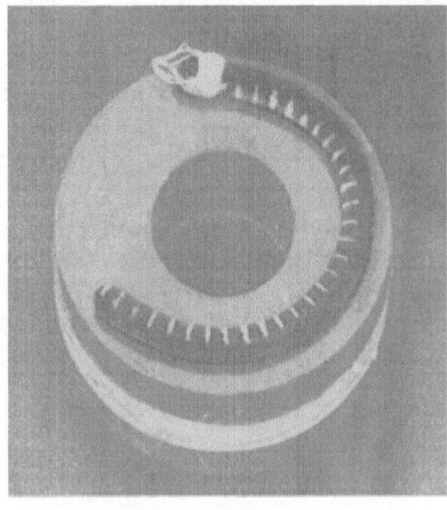

Fig. 3 Shape transition of ferrogels induced by non-uniform magnetic field

properly mixed the digitized picture coming from the video source with the computer screen image. By this method very short movements (one pixel on the screen) can be monitored and measured on the real time video image. The error of the measurement depends on the magnification, however, in our cases it was within 0.01 mm.

Two types of measurements have been performed by this equipment. The dependence of strain on the stress was determined by applying load in the absence of external magnetic field, and in the other case the magnetic field gradient was used to deform the gel, having no load.

Results and discussion

The effect of magnetic field on the shape of ferrogels as seen by naked eyes

In uniform magnetic field a ferrogel experiences no net force. When it is placed into a spatially non-uniform magnetic field, forces act on the magnetic particles, and the magnetic interactions are enhanced. The stronger field attracts the particles, and due to their small size and strong interactions with molecules of dispersing liquid and polymer chains they all move together. Changes in molecular conformation can accumulate and lead to shape changes. The principle of the ferrogel's shape transformation and motility lies in a unique magnetoelastic behavior. The magnetic field drives and controls the displacement and the final shape is set by the balance of magnetic and elastic interactions. The shape change can be bending, elongation, contraction and the combination of these basic distortions. This is demonstrated in Fig. 3. The magnetic field sensitive gels can be made to bend and straighten, as well as elongated and contracted repeatedly many times without damaging the gel. The response time to obtain the new equilibrium shape was found to be less than a second and seems to be independent of the size of the gel. It must be mentioned that all the shape changes reported here are completely reversible.

A peculiarity of ferrogel's shape distortion is the non-homogeneous deformation which can be seen in Figs. 3b and c. The gels were marked along their height by stripes

60

L. Barsi et al.
Gels with magnetic properties

perpendicular to the gel axis. The distance between subsequent stripes was constant. In non uniform magnetic field the displacement of subsequent stripes does not remain the same as in the case of homogeneous deformation; it varies along the gel axis, indicating the importance of nonhomogeneous shape distortion.

Magnetic properties of ferrogels

Although no work has been published in connection with the magnetic properties of ferrogels, there is reason to suppose that a significant similarity exists between the magnetic properties of ferrogels and that of ferrofluids [8]. According to this analogy the mono-domain ferromagnetic particles of colloidal size are the elementary carriers of a magnetic moment in the ferrogel. In absence of an applied field they are randomly oriented due to thermal agitation and thus the ferrogel has no net magnetization. As soon as an external field is applied, magnetic moments tend to align with the field to produce a bulk magnetic moment, M. With ordinary field strengths the tendency of the dipole moments to align with the applied field is partially overcome by thermal agitation. As the strength of field increases, all particles eventually align their moments along the direction of field and, as a result, the magnetization saturates. If the applied field is turned off, the particles quickly randomize, and M is again reduced to zero. Assuming the magnetization of an individual particle in the gel to be equal to the saturation magnetization of the pure ferromagnetic material, M_s the magnetization of a ferrogel, is predicted to be:

$$M = \phi_m M_s \mathscr{L}(\xi) = \phi_m M_s (\coth\xi - 1/\xi), \qquad (1)$$

where the parameter ξ of Langevin function, $\mathscr{L}(\xi)$ is defined as

$$\xi = \frac{\mu_0 m H}{k_B T}, \qquad (2)$$

where H represents the strength of an external magnetic field, μ_0 means the magnetic permeability of the vacuum, m is the magnetic moment of mono-domain ferromagnetic particles, ϕ_m stands for the volume fraction of magnetic particles in the whole gel and k_B denotes the Boltzmann constant.

According to Eq. (1) the magnetization of a ferrogel is in direct proportion to the concentration of ferromagnetic particles and their saturation magnetization. At small field strength a linear dependence between magnetization and field intensity is predicted, whereas at high field intensities the saturation magnetization is achieved.

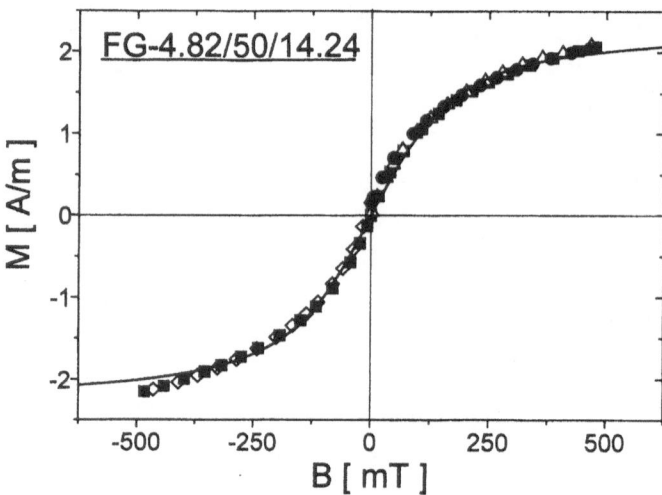

Fig. 4 Magnetization curve of a ferrogel measured at room temperature

On the basis of the Langevin type magnetization the initial magnetic susceptibility, χ obeys the Curie law:

$$\chi = \phi_m M_s \frac{\mu_0 m}{3k_B T}. \qquad (3)$$

Figure 4 shows the magnetization curve of a ferrogel. The magnetization is plotted against the magnetic induction, which is linearly proportional to the magnetic field strength. Two cycles have been measured and plotted. On the basis of experimental data, we can conclude that within the experimental accuracy, no hysteresis loop has been found. This means that the remanent magnetization of the ferrogel can be neglected. This finding is supported by other measurements performed on different ferrogels. This is an important result which says that in alternating magnetic field the transformation of magnetic energy into thermal energy is rather small. For this reason, devices made of ferrogels and subjected to alternating magnetic field – due to narrow hysteresis loop – are characterized by small energy loss per cycle.

On the basis of Fig. 4, we can also conclude that the Langevin approach can fit the experimental data, and a linear dependence between magnetization and field intensity can be used as a good approximation if the magnetic flux density $B < 120$ mT.

Elastic and magneto-elastic behavior of ferrogels

In the absence of an external magnetic field a ferrogel presents a mechanical behavior very close to that of a swollen filler-loaded network. Since a typical magnetic gel can be considered as a dilute magnetic system $\phi_m \le 0.1$ we may neglect the influence of magnetic interactions on the modulus. Thus the stress-strain dependence of a uni-directional deformed gel sample can be expressed on the basis of statistical theories [9, 10]

$$\sigma_n = G(\lambda - \lambda^{-2}) , \tag{4}$$

where the nominal stress σ_n is defined as the ratio of the equilibrium elastic force and the undeformed cross-sectional area of the sample. The deformation ratio λ is the length h (in the direction of force) divided by the corresponding undeformed length h_0. In Eq. (3) G stands for the modulus of ferrogel which can be expressed as a function of filler concentration

$$G = G_0(1 + k_E \phi_m) , \tag{5}$$

where G_0 denotes the modulus of gel without colloidal filler particles and k_E is the Einstein–Smallwood parameter. For non-interacting spherical particles $k_E = 2.5$. Equation (5) describes the reinforcement effect due to the filler-polymer interaction. The modulus of a ferrogel can be varied by the cross-linking density through G_0 and by the concentration of colloidal particles via ϕ_m.

It must be mentioned that, in many cases, Eq. (4) cannot be used to fit the experimental data. Deviation from the Gaussian theory of rubber elasticity may be due to finite chain extensibility and entanglement effects [9–11] and often represented by the Mooney plot:

$$\frac{\sigma_n}{\lambda - \lambda^{-2}} = C_1 + C_2 \lambda^{-1} , \tag{6}$$

where C_1 and C_2 are constants and the modulus is $G = C_1 + C_2$.

In order to test the applicability of Eqs. (5) and (6) for gels filled with magnetic particles, Mooney–Rivlin representation of experimental data obtained in the absence of external magnetic field is shown in Fig. 5. It may be seen that the nominal stress divided by $(\lambda - \lambda^{-2})$ seems to be slightly dependent on the strain. It is also seen that the modulus is strongly influenced by the cross-linking density.

In Fig. 6, we compare the experimental results of magneto-elastic investigations with that of the stress-strain measurements. This comparison is realized in the small deformation limit, where the Hooke law can be used as a good approximation. The relative displacement is plotted against the force. The experimental data show that there is no significant difference between two types of deformation. In both cases the stress-strain dependence can be described by a straight line, however, the slope of the straight line (modulus) obtained by mechanical stress was found to be smaller than that determined by non uniform magnetic field. This deviation may be attributed to the non-homogeneous deformation occurring in ferrogels.

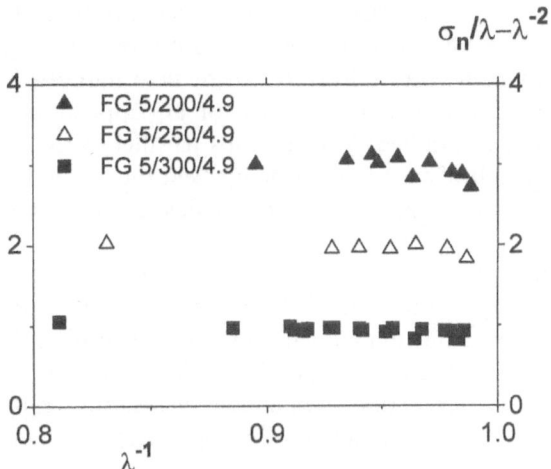

Fig. 5 Mooney–Rivlin representation of stress-strain data measured at elongation induced by mechanical stress

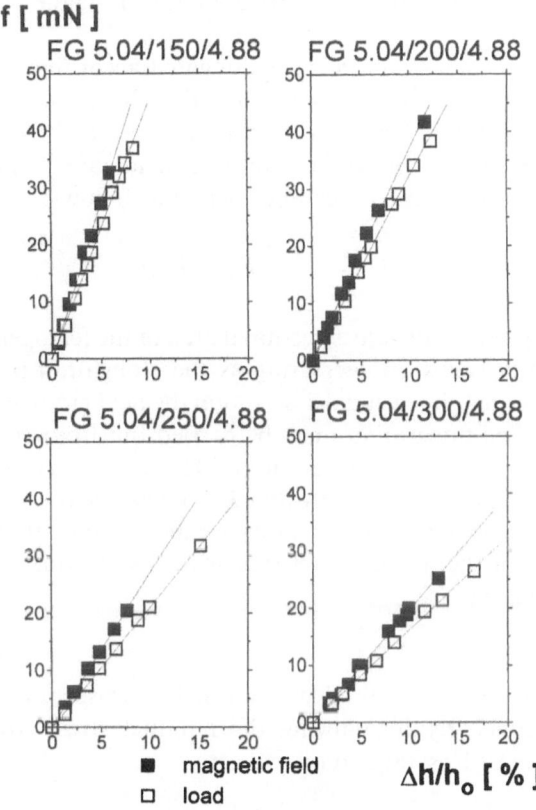

Fig. 6 Comparison of elastic and magneto elastic behavior of ferrogels at small deformation. □: mechanical experiment, ■: deformation was induced by non uniform external magnetic field

Theoretical interpretation of unidirectional magneto elastic behavior

Our description of magnetism occurring in ferrogels is based in part on the experimental fact that the ferromagnetic particles in the gel modifies the magnetic field produced by a current-carrying solenoid. The gel sets up its own magnetic field, which vectorially adds to the field induced by the electromagnet. In the presence of an applied field a magnetic field gradient ∇H develops parallel to the gel tube. As a result of an inhomogeneous field the gel undergoes a deformation and responds to the field gradient with a displacement.

The force measuring unit experiences a force given as the sum of weight and magnetic force developed under the action of a non-uniform magnetic field. By performing the experiments in a liquid having the same density as the gel, the measured force is due to magnetic interactions only. The magnetic force, f_M acting on the whole ferrogel cylinder can be given as

$$f_M = \int_V \mu_0(M\nabla)H dV , \tag{7}$$

where the integration must be carried out for the whole gel volume, V.

In the case of a cylindrical gel elongated according to Fig. 1 and taking into account that the elongation is induced by weak fields, we may assume a linear dependence between magnetization and field intensities ($M = \chi H$), therefore Eq. (7) can be written as follows:

$$f_M = a_s \int_{z_m - h}^{z_m} \mu_0 M\left(\frac{\partial M}{\partial z}\right) dz = a_s \frac{\mu_0\chi}{2}(H_h^2 - H_m^2) , \tag{8}$$

where a_s represents the cross-sectional area of the ferrogel. This quantity keeps on decreasing as the elongation increases. In a one-dimensional description the field gradient is ($\partial H/\partial z$). The integration must be carried out from the bottom $z_m - h$ to the top, z_m of the gel. The magnetic field strength at these points are H_h and H_m, respectively.

On the basis of our observations, we may assume that the volume of the magnetic gel remains constant during deformation, that is,

$$a_s = a_0 \cdot \lambda^{-1} , \tag{9}$$

where a_0 denotes the initial undeformed cross-sectional area of the gel. By introducing the nominal stress (by combination of Eqs. (8) and (9)), we get:

$$\sigma_n = -\frac{\mu_0\chi}{2\lambda}(H_h^2 - H_m^2) . \tag{10}$$

In equilibrium the stress due to magnetic interactions is counterbalanced by the network's elasticity. In general, the deformation induced by magnetic field cannot be considered as a homogeneous deformation, since $(M\nabla)H$ – the driving force – varies from point to point in space [12]. In our experiments, it was found that – due to the special distribution of magnetic field (Fig. 2) – the deviation from

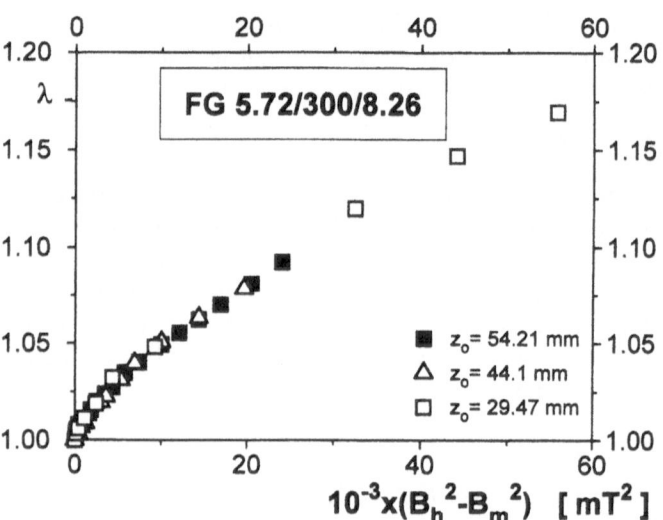

Fig. 7 Dependence of deformation ratio on the magnetic field strength measured at different initial position designated by Z_m. The solid line was fitted by Eq. (1) with parameter $M = 2.33[\coth(0.0145B) - \frac{1}{0.0145B}]$

homogeneous case is not significant, therefore the condition of equilibrium can be expressed as follows:

$$G(\lambda - \lambda^{-2}) - \frac{\mu_0\chi}{2\lambda}(H_h^2 - H_m^2) = 0 . \tag{11}$$

In this equation the Gaussian elasticity (Eq. (4)) is used.

Rearrangement of the above equation yields:

$$\lambda^3 - \beta(H_h^2 - H_m^2)\lambda - 1 = 0 , \tag{12}$$

where the parameter β is defined as

$$\beta = \frac{\mu_0\chi}{2G} , \tag{13}$$

and has a dimension of $m^2 A^{-2}$.

Equation (12) can be considered as a basic equation for describing the unidirectional magneto-elastic properties of ferrogels. It says if $H_h > H_m$, then $\lambda > 1$ and elongation occurs. For the opposite case, $H_h < H_m$, contraction is predicted.

It must be mentioned that the derivation of $\lambda(H)$ dependence on the basis of Eq. (12) is a rather difficult task, because λ and H_h are not independent quantities.

In order to check the applicability of Eq. (12) the dependence of λ against $B_h^2 - B_m^2$ is plotted in Fig. 7. In order to provide more evidence several experiments have been carried out on the same sample.

We have changed the axial position of the suspended gel tube by varying Z_m and thus the field strengths H_m and H_h (or B_m and B_h) as well as altering the field gradient inside the gel. It was found that no influence of the position of ferrogel on the λ-($B_h^2 - B_m^2$) dependence can be seen. All

Progr Colloid Polym Sci (1996) 102:57–63
© Steinkopff Verlag 1996

the measured points scatter around the same curve, supporting the validity of Eq. (13).

Summary

We have prepared magnetic field sensitive polymer gels. In the highly swollen hydrogels ferromagnetic particles of colloidal size are responsible for bulk magnetization induced by an external magnetic field. Elongation, contraction and bending can be realized by proper magnetic field distribution. We have studied the magnetization of ferrogels and it was found that no hysteresis loop occurs and the Langevin approach can be used to describe the magnetization phenomenon in ferrogels. An expression for the unidirectional magneto elastic properties has been derived and compared with experimental results.

Acknowledgements The authors would like to acknowledge the financial support of the Hungarian Academy of Sciences, OTKA T 015754.

References

1. Osada Y (1987) Advances in Polymer Science 82:1
2. De Rossi, Kawana K, Osada Y, Yamauchi A (Eds) (1991) Polymer Gels Fundamental and Biomedical Applications. Plenum Press, New York
3. Tanaka T (1982) Science 218:467
4. Zrínyi M, Barsi L, Büki A (1995) Europhysics Conference Abstract Vol 19, p 40
5. Zrínyi M, Barsi L, Büki A, Polymer Gels and Networks (in press)
6. Mark JE (1985) British Polymer Journal 17:144
7. Haas W, Zrínyi M, Kilian H-G, Heise B (1993) Colloid and Polymer Science 271:1024
8. Rosenweig RE (1985) Ferrohydrodynamics, Cambridge University Press, Cambridge, London, New York, New Rochelle, Melburbe, Sydney
9. Dusek K, Prins W (1969) Advances in Polymer Science 6:1
10. Tanaka T (1978) Phys Rev Lett 40:820
11. Mark JE, Erman B (1988) Rubberlike Elasticity a Molecular Primer. John Wiley Sons New York, Chichester, Brisbance, Toronto, Singapore
12. Kilian H-G (1987) Colloid and Polymer Science 265:410
13. Szabó D, Barsi L, Büki A, Zrinyi M, submitted to Models in Chemistry

Progr Colloid Polym Sci (1996) 102:64–70
Steinkopff Verlag 1996

P. Terech

Networks of surfactant-made physical organogels

Dr. P. Terech (✉)
Département de Recherche Fondamentale
sur la Matière Condensée
SI3M PCM
C.E.A.-Grenoble
17, rue des Martyrs
38054 Grenoble Cédex 09, France

Abstract Thermoreversible networks can be formed from associated small molecules in appropriate organic solvents. The present paper reports on the use of neutrons and x-ray scattering techniques to probe the structural features of the colloidal aggregates constituting the gel networks and on rheological experiments which characterize some of the dynamic and semi-static properties. Four types of gel systems, characterized by the nature of their junction zones, are discussed. Depending upon the chemical constitution of the gelators, fibers, semi-rigid molecular threads and rods can be formed and give, above the overlap threshold (of the order of 0.5–1%), gels with high yield stress values, pseudo-plastic fluids or thixotropic gels, respectively. The specific examples of 12-hydroxy-stearic acid (HSA), cholesteryl anthraquinone-2-carboxylate (CAQ), binuclear copper (II) tetracarboxylate (Cu_2S_8) and a trisubstituted metalloporphyrin, zinc (II) 5-(p-carboxyphenyl)-10, 15, 20-tris(p-hexadecyloxycarbonylphenyl) porphyrinate, (ZnP_3), are discussed.

Key words 61.12E-Neutron scattering techniques – 82.70D-Colloids

Introduction

Some low-molecular weight compounds can form gels in organic liquids. At a fixed temperature, in the gel domain of the phase diagram of these systems, the related three-dimensional networks can be either permanent or transient [1]. The corresponding nodes are more or less organized micro-domains which can involve crystallinity, lyotropism, transient ordering, and excluded volume interactions. The connecting chains, resulting from the aggregation process of the individual small gelator molecules, are usually rigid or semi-rigid.

The objective of the present paper is to demonstrate that a variety of networks can be observed in organogelator systems. The basic mechanical properties of the materials are investigated by rheology while the structural features of the gels are characterized by scattering techniques (using neutrons and x-rays). Four situations are discussed in this study and presented schematically in Scheme 1. Four representative examples of gelators (Scheme 2) are chosen so as to exhibit a large range of different molecular mechanisms involved in the aggregation reaction leading to different types of geometries of the fibrillar aggregates and different types of nodes in the three-dimensional solid-like networks.

Experimental

The gels were binary systems constituted of gelators dissolved at high temperature in organic liquids. On cooling, the viscoelastic materials were formed. The 0.1–15% wt concentration range was investigated.

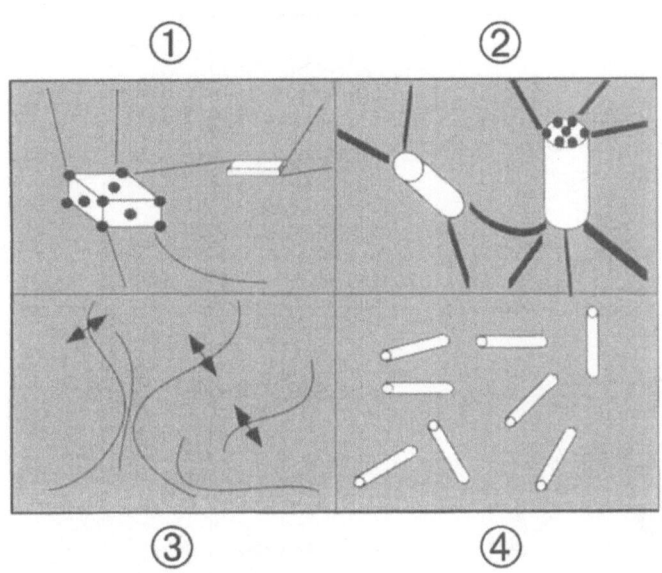

Scheme 1 A cartoon representation of the four types of nodes in the networks. I) crystalline; II) lyotropic; III) transient entanglements in "living polymers"; IV) long-range interactions

Synthesis and Sample Preparation

–HSA (12-hydroxystearic acid).

The optically active D-HSA compound was obtained as previously described [2]. The racemic DL-HSA and the solvents (benzene, cyclohexane, cis-decaline, tetralin, octane, nitrobenzene, hexafluorobenzene) were from Aldrich. Self-supporting gels were obtained, transparent in benzene and more turbid in alkanes. Xerogels were formed by slow evaporation of the solvent from the gels while attempts of crystallization in non-gelling solvents (methanol) gave only poor quality crystals justifying the use of powder samples for some WAXS studies.

–CAQ (cholesteryl anthraquinone-2-carboxylate).

The synthesis of CAQ has been described previously [3]. Due to the light sensitivity of the anthraquinonyl group, the yellowish CAQ powder was kept in the dark before use. Gelled solvents were dodecane, decane, 1-butanol, 1-octanol. The samples were self-supporting upon inversion of their sample holder and stable over a 3–4 day period before a solid-liquid phase separation was visually evident (gels in alcohols were the most stable).

–Cu_2S_8 (bicopper(II) tetracarboxylate complex).

The self-associating compound used in this study is a binuclear tetracarboxylate $Cu_2(O_2C-R)_4$ complex where R is a branched aliphatic chain ($R = CH(C_2H_5)C_4H_9$).

The abbreviation used for this molecule is Cu_2S_8 (S_8 refers to a complex with four substituted aliphatic chains of eight carbons each). Its synthesis has been described previously [4]. Convenient solvents were cyclohexane, $trans$-decalin and $tert$-butylcyclohexane (TBC). The samples were viscous and non-self-supporting materials described as jellies.

–ZnP_3 (Zinc(II) 5-(p-carboxyphenyl)-10,15,20-tris(p-hexadecyloxycarbonyl phenyl) porphyrinate).

The synthesis of ZnP_3 was obtained as a side product during the purification of the corresponding tetraester, Zinc(II) 5,10,15,20-tetrakis (p-hexadecyloxycarbonyl-phenyl) porphyrinate by column chromatography on basic alumina [5]. The porphyrin was characterized by mass spectroscopy, 1H NMR and UV-vis spectroscopy. In cyclohexane, a non-self-supporting jelly was obtained which was easily and reversibly converted to a liquid by simple shaking of the sample.

Rheological studies

The rheological measurements were made using a Rheometrics RMS800 and a Carri-Med CSL100 rheometers with a cone-plate geometry within the 0.001 to 100rd/s^{-1} angular frequency range. TBC was preferred over cyclohexane due to its high boiling point.

Scattering experiments

1 or 2 mm thickness quartz cells and 1 mm path-length cells with capton windows were used for neutron and x-ray experiments, respectively. Fully deuterated organic solvents (1-butanol, cyclohexane, benzene, etc. from Aldrich) and decane (from Cambridge) were used for the neutron scattering experiments. The momentum transfer $Q(\text{Å}^{-1})$ was defined as usual for pure elastic scattering and covered the average range $2.10^{-3}\text{Å}^{-1} < Q < 0.4\text{Å}^{-1}$. All spectra were recorded at room temperature.

SAXS data (for CAQ, ZnP_3) were obtained at the D.C.I. synchrotron source, LURE (Orsay, France) on the D22 spectrometer at a wavelength $\lambda = 1.458\text{Å}$ ($E = 8500 \text{ eV}$) and two distances (1.75 and 0.714 m).

The small-angle neutron scattering (SANS) measurements (for Cu_2S_8 and HSA) were made using the D11, D16, D17 spectrometers of the 58 MW reactor at ILL (Grenoble, France). Data were corrected for the empty beam signal and a light water standard was used to normalize the intensities. Additional data (CAQ) were obtained using the 8 m and 30 m spectrometers ($\lambda = 10\text{Å}$) of the 20 MW reactor at NIST (Gaithersburg, MD-USA). Absolute intensities were obtained by calibration with the

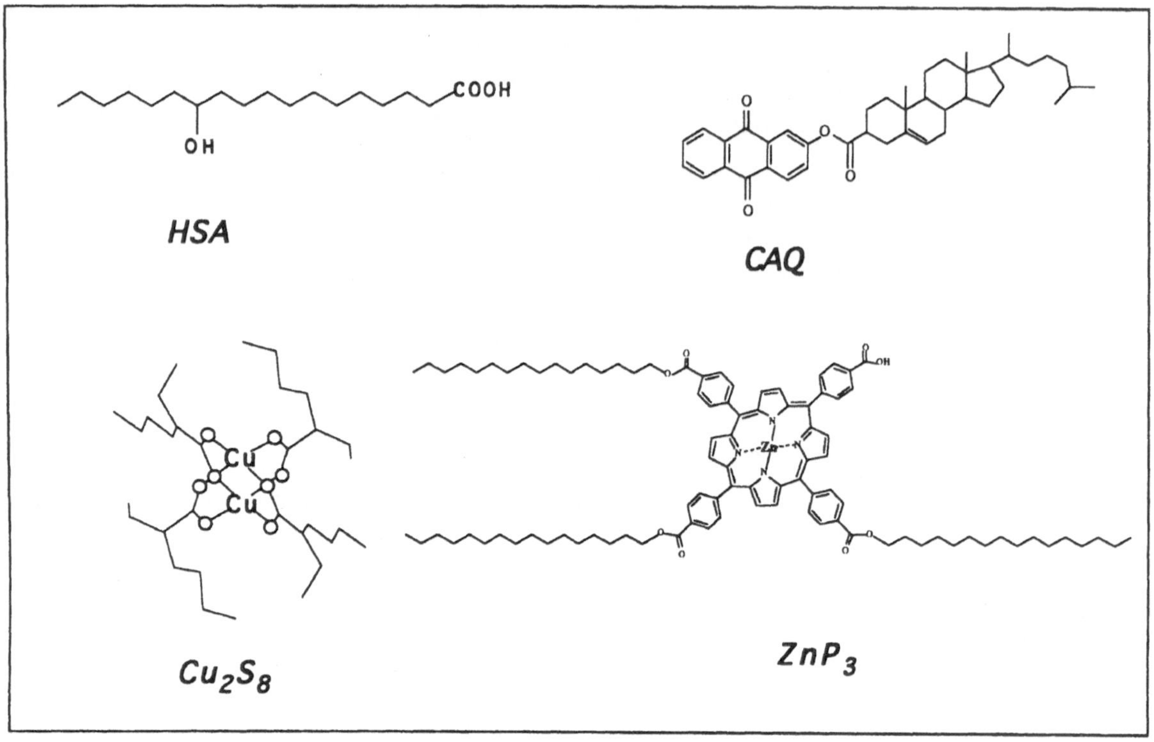

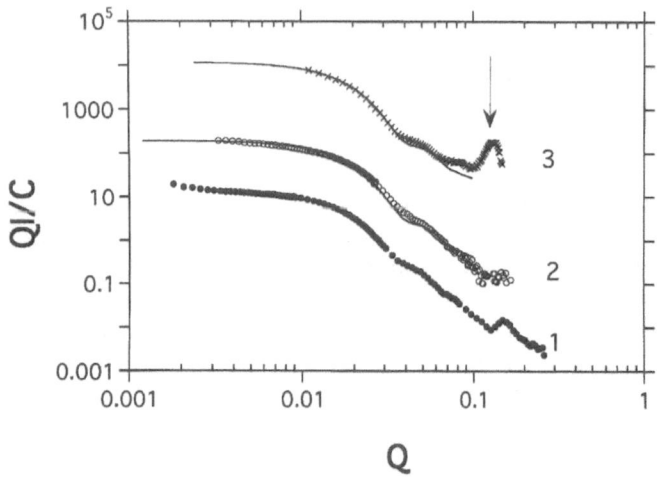

COOH

OH

HSA

CAQ

Cu

Cu

Cu₂S₈

ZnP₃

Scheme 2 Formuli of the organogelators corresponding to the four types of networks (see text). HSA: substituted fatty acid; CAQ: anthryl-appended steroid; Cu₂S₈: bicopper tetracarboxylate; ZnP₃: trisubstituted metalloporphyrin

Fig. 1 SANS of HSA gels in benzene. The cross-sectional intensity (QI) is plotted *versus* Q. 1) D-HSA, $C = 0.0235$ g·cm^{-3}, SANS; 2) DL-HSA $C = 0.0147$ g·cm^{-3}; 3) DL-HSA, $C = 0.030$ g·cm^{-3} (SAXS). The full lines are adjustments of Eq. (1) to the data ($a = 97$ Å, cross-sectional polydispersity $\varepsilon = 0.1$). The arrow indicates the (001) reflection

Results and analysis

Network 1: HSA

Figure 1 presents some SANS and SAXS data of HSA gels in benzene. The scattering behavior of the gels is not significantly altered by concentration and optical activity of HSA. The most significant features of the scattering consists of the Q^{-1} asymptotic low-angle plateau and a wide-angle Bragg peak ($Q \approx 0.14$ Å^{-1}). In addition, two oscillations of the intensity are observed in the intermediate Q-range located after the sharp intensity decay. The theoretical adjustments are made according to expressions (1) for SANS data [6, 7]. Comparable expressions are used for each of the fibrillar aggregates of the other networks 2, 3 and 4 (Scheme 1).

$$I(Q) = \frac{2C}{Q}\overline{\Delta b^2} M_L \int_0^{2\pi} \left[\frac{\sin(Qa\cos\varphi)}{Qa\cos\varphi} \cdot \frac{\sin(Qka\cos\varphi)}{Qka\cos\varphi} \right]^2 d\varphi$$

$$(1-1)$$

$$\overline{\Delta b} = b_2 - \rho_s v_2 \tag{1-2}$$

$$(Q\cdot I)_{Q\to 0} = (Q\cdot I)_0 \exp\left(-\frac{R_c^2}{2}Q^2 \right) \tag{1-3}$$

standard scatterings of a silica gel and a polystyrene sample for the 8 m and 30 m spectrometers, respectively. SANS data of ZnP₃ were obtained at the spectrometer PAXE at L.L.B. (Saclay, France).

Progr Colloid Polym Sci (1996) 102:64–70
© Steinkopff Verlag 1996

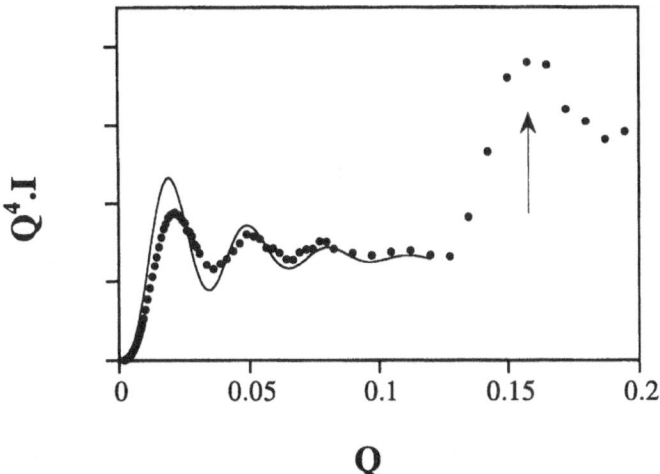

Fig. 2 Form-factor oscillations in the Porod region of a D-HOA/benzene gel ($C = 2.5\%$wt). The arrow indicates the Bragg reflection. Full line is an adjustment to Eq. (1) for a square cross-section: $k = 1$, $a = b = 107$ Å, $\varepsilon \approx 0.065$

where a, b are half the sides of the rectangular (or square) cross-section, k is its anisometry $k = b/a = B/A$ ($k = 1$ for a square, $k \to 0$ for lamellar shapes). C is the surfactant concentration in g·cm^{-3}, $\overline{\Delta b}$ is the specific neutron contrast (cm·g^{-1}) of the fiber, ρ_s is the neutron density of the solvent (cm·cm^{-3}). b_2 is the specific neutron density of the surfactant $b_2 = N_A \sum_{i=1}^{N_{at}} b_i/M$ where the summation is extended over the neutron scattering length values (b_i) of the N_{at} atoms of the gelator molecule. N_A is the Avogadro's number, M is the gelator molecular weight and v_2 is the specific volume of the surfactant in the fiber. The neutron contrasts of the racemic and chiral HSA derivatives are taken to be equal considering the related isotopic composition ($C_{18}H_{36}O_3$) and densities of the D and DL compounds [8]. For benzene gels of DL-HSA, a good fit to the experimental data is obtained for a square cross-sectional geometry and compares to that for the chiral benzene gels. Mean values using the "Guinier plots" $\ln QI$ *versus* Q^2 (expression 1–3), measured for dilute systems, are: number of aggregated HSA molecules per angström of linear fiber $n_L \approx 41.7$ mol·Å^{-1} and cross-sectional radius of gyration $R_c \approx 83.6$ Å giving $a = 102$ Å for fibers with a square cross-section. The geometry of the cross-section for HSA/benzene gels is confirmed in the Porod Q-region with the $Q^4 \cdot I$ vs Q plots (Fig. 2). The best fit to the position of the extrema is obtained for a square cross-section with 214 Å side.

The 3D solid-like network of HSA gels can be visualized as a random distribution of very long and rigid fibers interconnected by junction zones. Considering the concentration and solvent type effects on the scattering behavior ([2, 8] and reference cited therein), the shapes of the junction zones are crystalline bundles of fibers with an overall cross-sectional shape being square (in benzene) or more or less rectangular (in hexafluorobenzene, thickness ≈ 300 Å).

Rheology is used to probe during the kinetics of HSA aggregation the onset of an elastic modulus, G' (Fig. 3), whose typical value is 5.10^4 dyn·cm^{-2} (volume fraction $\phi = 1\%$). The plastic materials exhibit high yield stress values as is the case for viscoelastic solids with a permanent elasticity. At a temperature in the gel domain of the phase diagram, the nodes are permanent and "dry" crystalline-like microdomains as characterized by WAXS and DSC studies [2] on gels and xerogels. The internal molecular organization within the HSA ribbon is deduced from WAXS studies and from the analysis of the anisotropic scattering observed with oriented D-HSA/C_6F_6 gels [8]. The Bragg diffraction peak observed at $Q \approx 0.14$ Å^{-1} corresponds to the (001) reflection in a monoclinic crystallographic symmetry.

Fig. 3 Rheological time sweep experiments for DL-HSA/nitrobenzene gels. The warmed nitrobenzene solution ($C = 0.345\%$wt) was introduced between the cone and plate of the rheometer and cooled to cross the sol-gel threshold. Frequency = 1rad/s, strain = 1.02%. G': storage modulus, G'': loss modulus, T: temperature

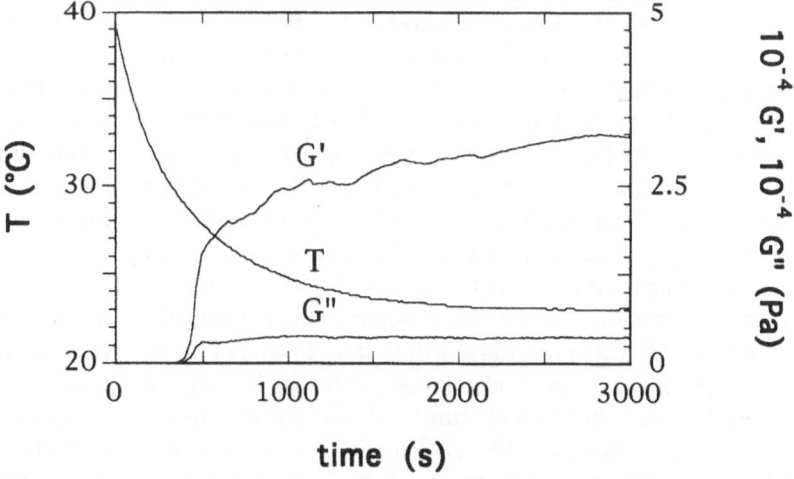

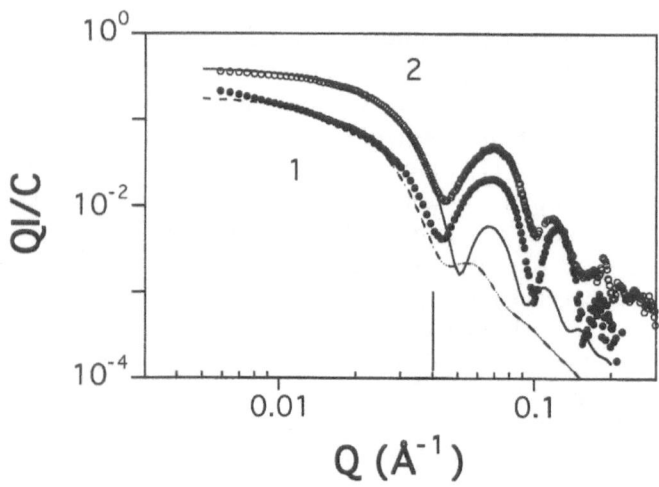

Fig. 4 SAXS cross-sectional intensity (QI/C) *versus* Q for CAQ organogels (logarithmic representation). (●), 1-octanol, $C = 0.0017$ g·cm^{-3}; 2: (○), decane, $C = 0.0043$ g·cm^{-3}. Lines are fits according to expression (1) : 1, full line, $R_0 = 72$ Å, $\varepsilon = 0.1$; 2, dotted line, $R_0 = 75$ Å, $\varepsilon = 0.2$. The vertical bar is a visual separator between the low-angle and large-angle parts of the scattering

Network 2: CAQ

SAXS data of CAQ gels in octanol and decane are shown in Fig. 4 which reveals that the connecting aggregates are very long and fully rigid fibers. The analysis of the low-angle part of the scattering [9, 10] shows that the cross-sectional symmetry is dependent upon the solvent type (circular in decane, rectangular in 1-octanol). Typical values of the geometrical parameters are: length $\rightarrow \infty$, diameter ~ 102 Å. The scattering curves mix the form-factor and the structure factors of the CAQ aggregates. The interpretation of the large-angle diffraction pattern assumes that the nodes are lyotropic microdomains whose symmetry evolves from hexagonal ($p6m$) in decane to lamellar-like in 1-octanol. The analysis is consistent with the observation of Schlieren textures in 1-octanol gels [9].

CAQ is a member of a family of steroid-based anthryl-appended organogelators [3, 11] in which each part (steroid, anthryl and functionalized link) can play a determinant role in the aggregation process. For CAQ, the anthryl part has been shown to master the molecular association. Electronic microscopy of a close derivative of this family [3] confirms the existence of ribbons (which can be helical depending on the polarity of the solvent) whose cross-sectional dimensions are comparable to those observed by SAS. CAQ organogels in decane consist of fibers interacting in lyotropic microdomains such as bundles of at least three hexagonal units of cylindrical fibers ($R_0 \approx 51$ Å). In alcohols, the local molecular organization is the same but the arrangement of the "junction zones"

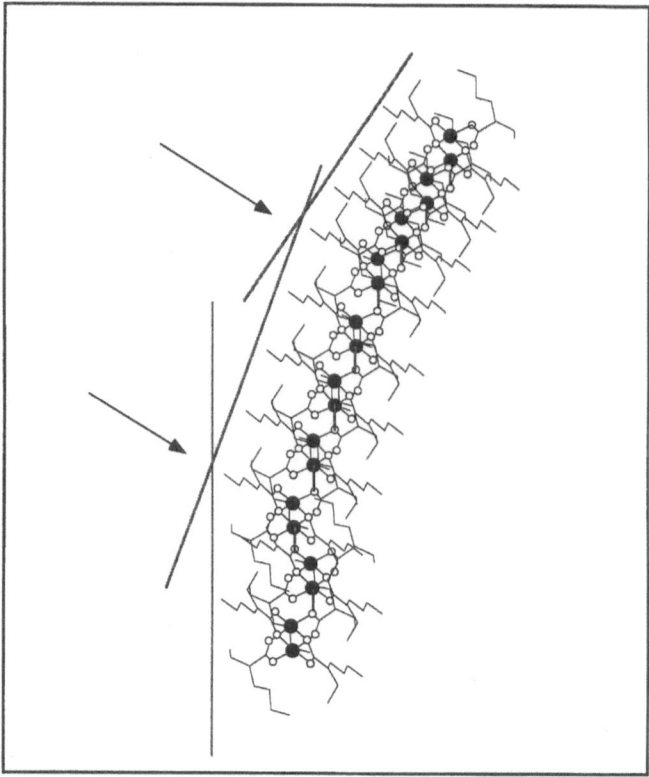

Fig. 5 Unidirectional aggregate made up of Cu_2S_8 molecules connected by copper-oxygen bonds. The copper-copper spacings in the assembly are similar either to those in the hexagonal discotic mesophase or to those in the layered crystalline solid. Arrows indicate a curvature change

results from a merging of more rectangular fibers (ribbons) in lamellar-like structures.

Network 3: Cu_2S_8

SANS measurements [12] show that the aggregates are "molecular threads" (with only one molecule per cross-section, Fig. 5). The chain is very thin (diameter ca. 17 Å) and semi-flexible. The molecular stacking involves copper–oxygen coordination bonds and the statistical length results from a thermal equilibrium with the bulk solution.

Rheological experiments [12–15] reveal that the material is a viscoelastic liquid (pseudo-plastic) exhibiting a stress relaxation (Fig. 6) which is not far from being monoexponential in appropriate concentration conditions in the semi-dilute regime. These features are typical of the so-called "living polymers" [16–18]. The network is transient and the chains undergo scission/recombination reactions. The materials exhibit semi-static and dynamic relationships (diffusion coefficient, plateau modulus, terminal relaxation time, etc.) typical of some polymeric

Progr Colloid Polym Sci (1996) 102:64–70
© Steinkopff Verlag 1996

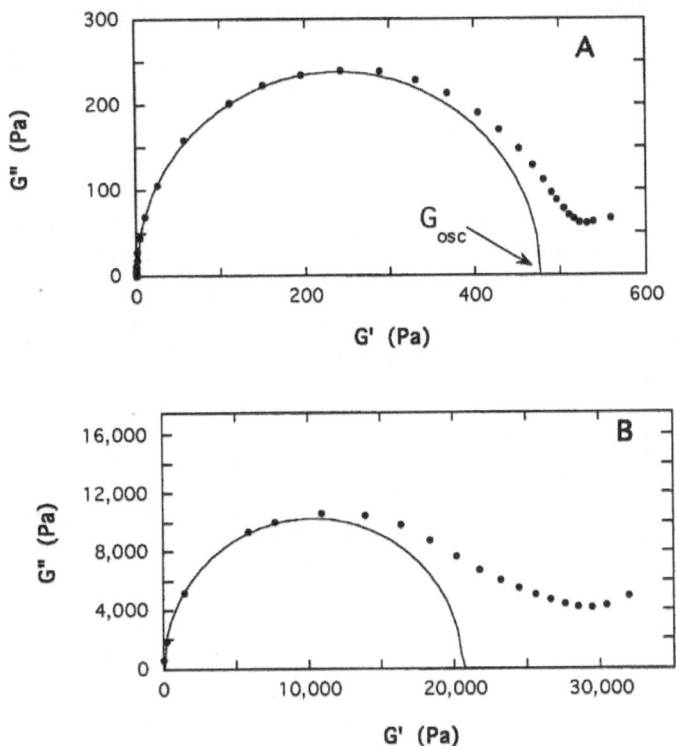

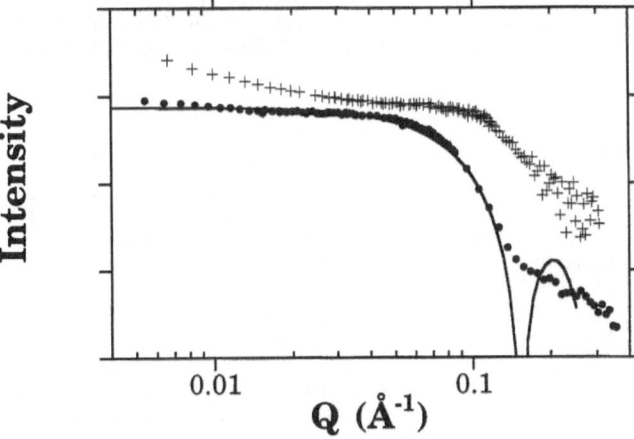

Fig. 7 SAS of ZnP$_3$ jellies in cyclohexane ($C \approx 2\%$ wt). (+) SAXS; (●) SANS. Full line is an adjustment for an homogeneous rod with a diameter 50 Å ($\varepsilon = 0$)

Fig. 6 Cole-Cole plots for two viscoelastic Cu$_2$S$_8$ solutions (A: $\phi = 0.71\%$; B: $\phi = 2.2\%$). The semi-circles are the corresponding adjustments for monoexponential Maxwellian stress relaxation (osculating semi-circles G_{osc})

systems and scission kinetics typical of surfactant-made chains. The scission is a monomolecular reaction and the contact zones involve transient ordered microdomains (in the continuity of the phase diagram where discotic, hexagonal and nematic domains are found) accounting for the elastic modulus at high frequency ($G' \sim 120\,\text{Pa}$ for $\phi \approx 1\%$). The flexibility arises from the combination of crankshaft-like (discotic-like) and layered (solid-like) sequences of connection sites in the aggregation reaction. The binding energy ($\sim 16\,kT$ for copper/oxygen) and the chain flexibility determine the special dynamical properties of the materials and some departures from the theoretical behavior expected for ideal "living polymers" can be understood in the context of the motion of chain extremities in the tight nets of concentrated Cu$_2$S$_8$ solutions. Transient interactions between different chains act as entanglements in the system while the role of some chain terminators (water, alcohols) in these microdomains has to be clarified.

Network 4: ZnP$_3$

Initial SANS and SAXS experiments on this novel gelator system [5] reveal the existence of rod-like aggregates

(Fig. 7). Contrast variations between the two radiations demonstrates the existence of a "molecular thread" organization obtained through a stacking of metalloporphyrins. The rods have a metallic core and a shell (diameter ~ 50 Å) of high densities of π electrons. Further characterizations of the mechanical properties (thixotropy of a system with a low yield value) and of the optical properties (exciton coupling in the molecular assembly) are currently in progress. Scattering evidence of the existence of finite length rods (which are easily orientable by shear stresses) have been obtained, thus suggesting that the network could result from long-range interactions between the rods. A tentative and speculative scheme of the network is presented in (Scheme 1).

Conclusion

The above described examples of the organogel networks made of low-molecular weight compounds demonstrate the great variety of structural and rheological behaviors which can be observed in these systems. Depending upon the chemical functionality of the gelators, hydrogen bonding, organometallic coordination bonding, partial recovering of polarized aromatic systems can be the driving mechanisms for the aggregation reaction. Unidirectional aggregates are always observed in these systems. Depending on the interfacial structure of the fibers, various interactions can develop between adjacent fibers in the nodes (or junction zones) of the networks. Crystalline, lyotropic (swollen) microdomains, transient associations and excluded volume interactions are respectively acting as nodes in the networks 1,2,3,4 of Scheme 1.

Acknowledgments. The ILL, LLB, LURE, NIST facilities are acknowledged for providing the neutron and all technical supports. Prof. R.G. Weiss and Drs I. Furman, G. Gebel, E. Ostuni, P. Maldivi, R. Ramasseul, V. Rodriguez are acknowledged for their contributions to this work.

References

1. Terech P (1996) In: Low-molecular weight organogelators, Specialist surfactants, Blackie Academic & Professional, Chapman & Hall (in press)
2. Terech P, Rodriguez V, Barnes J, McKenna GB (1994) Langmuir 10:3406–3418
3. Lin Y-C, Kachar B, Weiss RG (1989) J Am Chem Soc 111:5542–5551
4. Maldivi P (1989) Thesis, Grenoble, France
5. Terech P, Gebel G, Ramasseul R, Langmuir (in press)
6. Cabane B (1987) Surfactant solutions, Surfactant Science Series 22:57, Zana R (ed) Marcel Dekker Inc, New York
7. Glatter O, Kratky O (1982) In: Small angle X-ray scattering, Academic Press, London
8. Terech P (1992) J Phys II France 2:2181–2195
9. Terech P, Ostuni E, Weiss RG (1996) J Phys Chem 100:3759–3766
10. Terech P, Furman I, Weiss RG (1995) J Phys Chem 99:9558–9566
11. Mukkamala R, Weiss RG (1995) J Chem Soc, Chem Commun, pp 375–376
12. Terech P, Schaffhauser V, Maldivi P, Guenet JM (1992) Europhys Lett 17: 515–521
13. Terech P, Schaffhauser V, Maldivi P, Guenet JM (1992) Langmuir 8: 2104–2106
14. Terech P, Maldivi P, Dammer C (1994) J Phys France II 4:1799–1811
15. Dammer C, Maldivi P, Terech P, Guenet JM (1995) Langmuir 11: 1500–1506
16. Cates ME (1987) Macromolecules 20:2289–2296
17. Cates ME (1988) J Phys France 49: 1593–1600
18. Cates ME, Candau SJ (1990) J Phys Condens Matter 2: 6869–6892

Progr Colloid Polym Sci (1996) 102:71–75
© Steinkopff Verlag 1996

P. Dokić
I. Sefer
V. Sovilj

Shear-induced polymer interactions and gelation in urea-formaldehyde polycondensates

Prof. Dr. P. Dokić (✉) · I. Sefer · V. Sovilj
University of Novi Sad
Tehnoloski fakultet
Buleavr cara Lazara 1
21000 Novi Sad, Yugoslavia

Abstract Urea and formaldehyde polycondensation was performed in shear and non shear condition under different pH values (3.5, 4, 4.5) and amounts of acetate buffer (0.14%, 0.41%, 0.69%). The reaction was ceased by changing pH from acidic to basic one when gel point was reached. This polycondensate represents a medium concentrated dispersion of polymer clusters in aqueos solution of short polymer chains and the buffer. Structural changes occurring within the reacting system were followed by rheological measurements and chemical analysis. The flow type was a plastic one. The dependence of gelation time (gel point) upon rate of shear was determined.

The concentration of the dispersion was further increased in the evaporation process under vacuum and small rate of shear. High concentration products obtained (concentration varied from 60–92%) were subjected to the shear force of varying intensity (shear rate up to $20\,000\ \text{s}^{-1}$) which induced polymer chain interaction and gelation in shear. Due to various factors, very complex rheological behaviour comprising shear thinning, rheopexy (antithixotropy) and thixotropy was obtained. The type and mechanism of structural linking in the process of building up of the gel network could be deduced.

Key words Gelation in shear – gel – rheology – thixotropy – rheopexy

Introduction

Recently, gelation of aqueous macromolecular solutions, the proces in which under rather complex mechanism a liquid solution is transformed into an elastic or thixotropic solid, has received great attention. Several excellent review papers dealing with classification of various types of gels and different mechanisms of gelation [1–3], as well as with recognition and understanding of the subprocesses encountered in the sol–gel transition [4] have been published.

Gelation, either chemical or physical, is the process leading to dramatic changes of the flow properties of the system. The gel once formed can undergo further changes in the internal structure, i.e., to an aging process which can be followed rheologically [5].

In following gelation process by rheological measurements it is undesirable to disturb building up of internal gel structure by mechanical force applied. On the other hand, there is a possibility to induce and influence gelation of the macro-molecular system subjecting it to a shear force of different intensity [6, 7]. Gelation under shear conditions seems to offer the facts which can be used in explaining the mechanisms and factors governing building up of supramolecular structure of the gel formed in either chemical or physical processes.

Urea and formaldehyde react in an acidic aqueous solution to give linear and branched polymers of different size and structure. Long chain branching starts as soon as the number of urea units in the linear chain is greater than five [8], giving branched polymer clusters. The clusters formed together with linear chains can crosslink to form a sort of supramolecular structure and three-dimensional network, i.e., finally a solid gel. Not only individual polymers are changing in size and shape but the system they build is changing with reaction time. Besides the time, the structure and the properties of the system are influenced by various factors such as urea–formaldehyde ratio, pH value, amount of buffer added, temperature, etc. [9]. Structure changes within the system can be followed rheologically. In that way the time gelation starts (gel point) can be detected and polycondensation ceases at that moment. If the polycondensation is stopped at the gel point, properties of the system could be further changed by applying a shear force. This fact gives a possibility to study gelation of the system under shear conditions.

Experimental

In this investigation urea–formaldehyde polycondensation was performed in shear and non shear conditions. The reaction was ceased by changing pH from acidic to basic one when gel point was reached.

The reaction of condensataion of urea and formaldehyde (molar ratio 1:2) was performed under different pH values (3.5, 4, 4.5) and amount of acetate buffer (0.14%, 0.41%, 0.69%) at the temperature of 62 °C. The reaction was ceased by adjusting pH to 7 and urea/formaldehyde ratio to 1:1.5 (to neutralize free formaldehyde). Reaction time was up to 280 min when the reaction was in shear (D: 0.54, 1.08, 2.16, 9.73, 14.9 and 21.6 s^{-1}) and up to 120 min in non shear conditions.

The polycondensate (dry content 45.3 − 52.5%) was submitted to water evaporation at 60 °C under vacuum (10660 Pa in a rotary evaporator) in order to get more concentrated products (concentration range 55–92.5%). Dry content was determined by drying at 105 °C for 5 h and refractometrically. These products were subjected to a rather intense steady shear (shear rate up to 20 000 s^{-1}) in the measuring device of the rotational viscometer in order to study shear induced gelation and linear chain and cluster interactions. Rheological measurements were performed by the rotational viscometer equipped with coaxial cylinders and plate and cone measuring devices (Haake, Germany). Shear stress (τ) was determined in dependence on shear rate (D) by continuous loop method at a constant rotor acceleration.

Results

Rheological characteristics of urea formaldehyde reacting system

During the polycondensation reaction rheological characteristics were determined continuously in steady shear or in regular time intervals in non shear conditions. In this way shear stress, i.e., viscosity dependences on reaction time and flow curves were obtained.

In steady shear conditions shear stress and apparent viscosity were increasing with the reaction time. The viscosity (η_a) – time (t) dependence was influenced by the intensity of shear (Fig. 1). It can be seen from Fig. 1 that the lower the value of shear rate the higher the viscosity in the whole interval of reaction time. After some specific time elapsed there was rather a steep increase in the viscosity. Some limiting value of the time, t_g (min) corresponding to the infinitively high viscosity ($\eta_a \to \infty$) can be deduced from the η_a – t curves. When an appropriate function (giving $\eta_a = 0$ for $t = 0$ and $\eta \to \infty$ for $t = t_g$) was used the time t_g was determined. This time can be taken as the beginning of gelation (the gel point) due to polymer clusters crosslinking. The gelation time is dependent on the shear rate used. When D is 0.541, 1.082 and 2.164 s^{-1}, t_g is

Fig. 1 The changes in the viscosity of urea–formaldehyde polycondensate with the reaction time

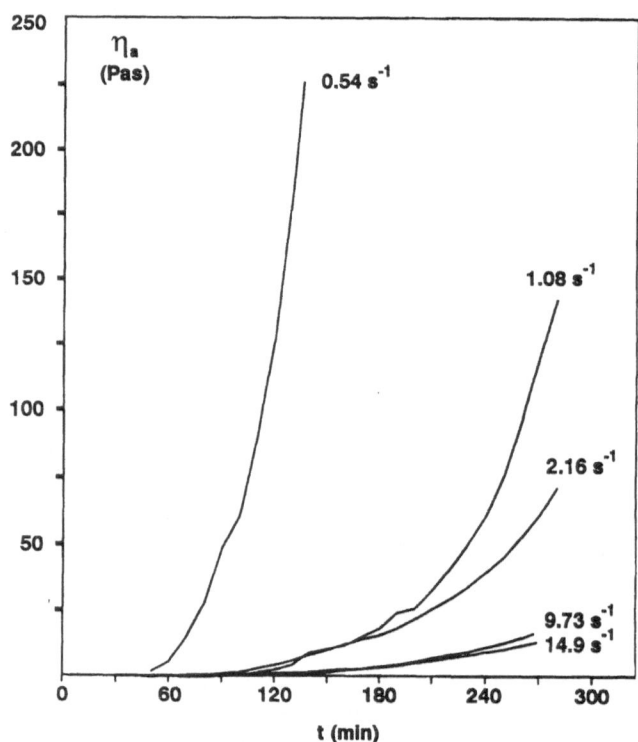

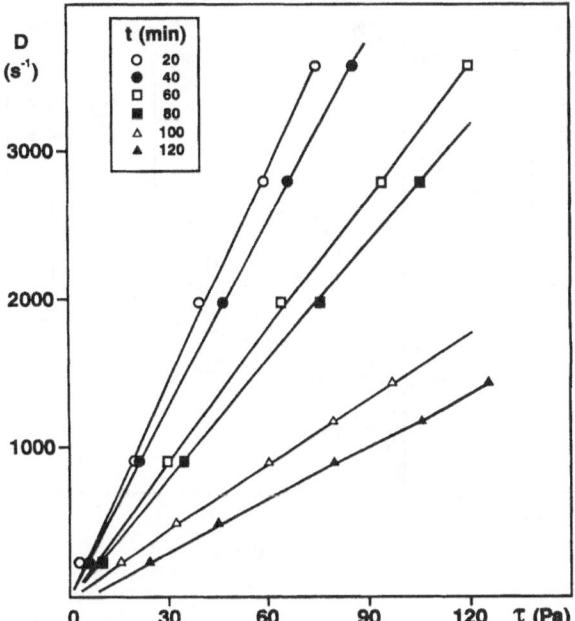

Fig. 2 Flow curves of urea–formaldehyde polycondensates of different reaction times

Table 1 The values of plastic viscosity η_{pl} (mPas) of urea–formaldehyde polycondensates of different pH and amount of acetate buffer in dependence on reaction time

pH	Amount of buffer (%)	Reaction time (min)					
		20	40	60	80	100	120
	0.14	140	208	350	880	–	–
3.5	0.41	28	29	49	62	–	95
	0.69	26	45	59	62	43	90
	0.14	28	28	31	41	59	130
4.0	0.41	21	24	34	38	68	90
	0.69	19	24	28	–	50	91
	0.14	7	15	19	22	–	22
4.5	0.41	25	20	22	29	43	56
	0.69	16	16	17	32	39	49

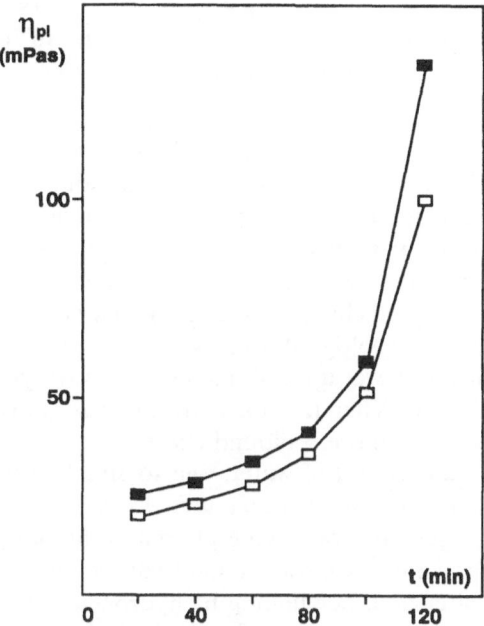

Fig. 3 The changes of plastic viscosity with the reaction time in non-shear conditions, (■) – 0.14% and (□) – 0.69% of buffer

156, 280 and 356 min respectively. The dependence of the time t_g on the shear rate is:

$$t_g = t_{go} + A \cdot D , \tag{1}$$

where t_{go} is gelation time when $D = 0$, i.e., in non shear conditions and A is a constant. For the system examined $t_{go} = 120$ min and $A = 113$. That means, if the polycondensation took place in non shear conditons the expected gelation time (gel point) would be 120 min.

In non shear conditons flow curve was determined every 20 min (Fig. 2). Shear stress – shear rate dependences are linear ones with small intercept on the τ axes, i.e., the flow type is Bingham plastic. Yield stress and plastic viscosity were calculated and their dependences on pH, amount of buffer (Table 1) and dry content were found. Changing of pH value from acidic to basic one and formaldehyde neutralization did not change the flow type.

Plastic viscosities, as can be seen from Table 1, increase with the reaction time (Fig. 3) by the dependence type similar to the one obtained in steady shear conditions. Figure 3 shows the gel point is about 120 min, which is in good agreement with the value obtained from Eq. (1) for the zero shear rate.

Flow curves and plastic viscosities from Table 1 together with solubility determinataions and microscopic examinations of the polycondensates allow to conclude the system is a medium concentrated dispersion (dry content 45.3–52.5%) of polymer clusters entangled and spherical in shape in aqueous solution of short polymer chains and the

buffer. Further physical treatment of the former dispersion can induce gelation and produce the gel by interconnecting polymer clusters.

Rheological behaviour of concentrated polycondensates – gelation in shear

In the evaporation process, where small shear force was acting, more concentrated polycondensates were obtained with the concentration ranging from 52.5 to 92%. These polycondensates were submitted to rather intense shear (rate of shear up to $20\,000\ s^{-1}$) within the viscometer measuring device. In fact, shear rate was increasing

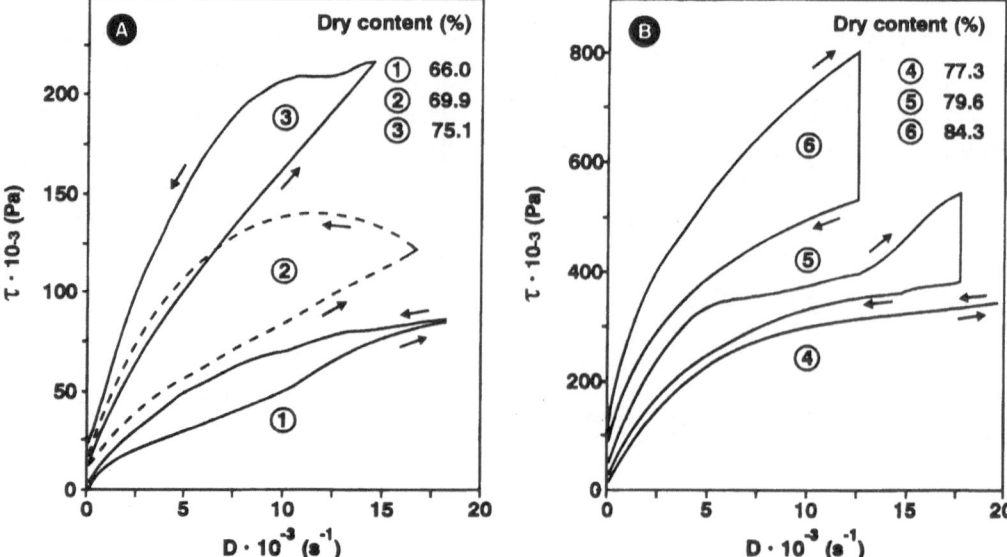

Fig. 4 Flow curves of concentrated urea–formaldehyde polycondensates (A) rheopexy and (B) thixotropy and shear thinning

continuously from zero up to some maximal value (depending of viscometer measuring range) and after reaching maximum shear rate was decreased until zero. In this way, up and down flow curves were obtained. These flow curves reflect internal structural changes which took place during the evaporation and rheological process.

If the concentration was up to about 60% the flow type still was a plastic one. When the concentration was above 60% shear force has a more profound effect.

During the evaporation of water, due to small shear force and decreasing distance between urea–formaldehyde polymer clusters, gelation was taking place and the polymer network was set up by covalent and hydrogen bonding, including water molecule linking by hydrogen bonds and polymer chains entanglement. Covalent bonds can be established between two CH_2–OH groups (leading to two types of intermolecular bridges) or $\diagdown$NH and –CH_2–OH groups of neighbour clusters and chains. Hydrogen bonds are set up between $\diagdown$NH and O=C$\diagdown$, –CH_2OH and O=C$\diagdown$, and two –CH_2OH groups, including solvation by water molecules and intermolecular linking over water. The greater the polymer concentration the higher the amount of the links set up and the degree of building up of the gel network. Depending on the amount of the links established various flow types were exhibited.

In the rheological process the phenomena taking place in the evaporation are continued and increased by steady shear of high intensity (D up to $20\,000\ s^{-1}$) and, combined with an orientation effect, it brought about rather complex gelation. In fact, the gelation was the result of two simultaneous resultant processes, i.e., setting up and disruption of structural links. All the former factors gave rise to a very

complex rheological behaviour (Fig. 4) comprising shear thinning, rheopexy (antithixotropy) and thixotropy.

In the concentration range from about 60 to 76%, up and down flow curves indicate (Fig. 4A) at smaller shear rates an orientation of polymer chains and kinetic aggregation of polymer clusters which were followed by oriented chains interaction and linking and clusters interlinking at higher shear rates. Down flow curves show higher shear stress and viscosity than up curves due to shear induced crosslinking mentioned and building up of network structure. At the concentration of 77.3% there is obviously a balance between structure making and structure breaking and the flow is pseudoplastic. In the concentration range above about 79% (Fig. 4B) the degree of setting up (which occurred in the evaporation process) of the polymer network is the highest and the system is a thixotropic one, i.e., a thixotropic gel. During the flow of these systems network disruption prevails over network building.

When all the concentrated polycondensates were diluted by adding water until the concentration of 66% was obtained and they were subject to flow afterwards all of them showed rheopexy.

If all the rheopectic loops of the diluted systems (66%) were compared among themselves a greater loop area was noticed for the samples diluted from the concentrations closer to 66%. Besides, the viscosity of diluted systems (66%) was higher than the visocsty of the original 66% polycondensate whether the concentration is the same or not. These facts show that during the gelation in the evaporation process covalent bonds were established which are not interrupted during the network disturbance

Progr Colloid Polym Sci (1996) 102:71–75
© Steinkopff Verlag 1996

and disruption by the dilution. Because they cannot be interrupted and because their number is greater than the diluted samples obtained from higher concentration, the possibility for the network building in shear is diminished compared to the original 66% polycondensate.

The rheological behaviour of various polycondensates examined show the gelation during the evaporation is characterized by the network build-up from polymer clusters (particles) and polymer streched and entangled chains interconnected by covalent and hydrogen bonding and water linking by hydrogen bonds. When the gelation is induced and governed by shear force hydrogen bonding combined with the orientation and kinetic aggregation effects is responsible for the network set up.

References

1. Clark A, Ross-Murphy B (1987) Adv Polymer Sci 83:57
2. Russo S (1987) Reversible Polymeric Gels and Related Systems. ACS Symposium Series 350, New York
3. Burchard W, Ross-Murphy B (1988) Physical Networks: Polymers and Gels. Elsevier Applied Science, London & New York
4. Djabourov M (1991) Polymer International 25:135–143
5. Djaković Lj, Dokić P (1972) Die Stärke 24:195–201
6. Djaković Lj, Dokić P, Kabić B (1972) Hemijska industrija – Chemical Industry (Beograd) 26:200–205
7. Carralho W, Djaburov M, Denis A, Schoentjes M (1994) In: Gallegos C (ed) Progress and Trends in Rheology IV. Dietrich Steinkopff Verlag, Darmstadt, p 449
8. Beachem T (1963) J Org Chem 28:1876
9. Djaković Lj, Dokić P, Šefer I (1983) Book of abstracts of VIIth Yugoslav congress on chemistry and chemical technology. Novi Sad (Yugoslavia), p V-49

Progr Colloid Polym Sci (1996) 102:76–81
© Steinkopff Verlag 1996

L. Halász
O. Vorster

Gelation in reactive polyester powder coating systems

Dr. L. Halász (✉) · O. Vorster
Technical University of Budapest
Department of Physical Chemistry
1521 Budapest, Hungary

Abstract The curing process of carboxyl terminated polyester – triglycidyl cianurate system was investigated by thermoanalytical and rheological methods. On the basis of evaluation of activation energies the curing process was divided into three stages and the reaction rates were calculated for these stages. Simplified kinetics equations were derived for the three stages of process and the feasibility of these equations was evaluated. A simple rheological evaluation method was elaborated to calculate the rate constants from the measured rheological parameters. The method gave good results.

Key words Epoxy – curing – gel – kinetics – viscoelasticity

Introduction

The reactive polyester powder coating technology plays an important role in painting techniques. The main reason for this is the excellent weather resistance of polyester powder coating systems cured with triglycidyl isocyanurate (TGIC).

The reaction between epoxy and acid functional groups is the most important curing reaction used in practice for crosslinking of thermosetting powder coatings. Schechter et al. [1–3] anticipated the following four possible reactions in the curing of epoxy resins with polybasic acids:

● Ring opening addition of the acid group to the epoxy group, leading to the formation of the corresponding hydroxyl ester:

$$RCOOH + CH_2\text{--}CHR_1 \rightarrow RCOOCH_2 \underset{\overset{|}{OH}}{CHR_1} \qquad [1]$$
$$\diagdown O \diagup$$

● Esterification between the acid and the hydroxyl group formed in the above-mentioned reaction:

$$RCOOH + RCOOCH_2\underset{\overset{|}{OH}}{CHR_1} \rightarrow RCOOCH_2 \underset{\overset{|}{OCOR}}{CHR_1} + H_2O \qquad [2]$$

● Ring opening addition of the hydroxyl group formed in reaction (1) to the epoxy group, resulting in the corresponding ether alcohol:

$$[3]$$

$$RCOOCH_2\underset{\overset{|}{OH}}{CHR_1} + CH_2\text{--}CHR_1 \rightarrow RCOOCH_2 \underset{\overset{|}{OCH_2CHR_1OH}}{CHR_1}$$
$$\diagdown O \diagup$$

● Hydrolysis of the epoxy ring by water:

$$H_2O + CH_2\text{--}CHR_1 \rightarrow HOCH_2\underset{\overset{|}{OH}}{CHR_1} . \qquad [4]$$
$$\diagdown O \diagup$$

The extent to which the above-mentioned reactions take place depends on the catalysts used in the system. In the absence of catalyst the rates at which the first two reactions proceed are comparable [4]. The water released in the condensation reaction volatilise from the film during the curing process. Reaction [4] may take place when the resin matrix or powder coating is contaminated by moisture. No mention has been made of the sequence or of the parallel completion of reactions.

Assuming that the whole curing process is a single reaction with single activation energy, then if the Arrhenius law is fulfilled the reaction rate can be written as [5]:

$$\frac{d\alpha}{dt} = kg(\alpha) = k_0 g(\alpha) \exp\frac{-E}{RT} , \qquad (1)$$

where α is the extent of the reaction, k is the rate constant, assumed to depend on the temperature in accordance with the Arrhenius law, $g(\alpha)$ is a function of the extent of the reaction, E is the activation energy, R is the universal gas constant, and T is the curing temperature in Kelvin. Considering that for a given extent of reaction, $g(\alpha)$ takes the same form regardless of curing temperature, it is possible to determine the activation energy for a given extent of reaction without knowing the form of $g(\alpha)$. Rearranging Eq. (1) and integrating it between a curing time $t = 0$ where $\alpha = 0$ and a time t with an extent of reaction of α, and taking logarithm, for a given extent of reaction, at constant temperature we obtain the following:

$$\ln t = A + \frac{E}{RT} . \qquad (2)$$

For each degree of conversion, A is a constant that has the following value:

$$A = \ln\left(\int_0^\alpha \frac{d\alpha}{g(\alpha)}\right) - \ln k_0 . \qquad (3)$$

Equation (2) represents a linear relationship between the logarithm of time necessary to reach a given extent of reaction and the inverse of the curing temperature. By applying Eq. (2) at a series of temperatures it is possible to determine the activation energy for the different extents of reaction from the slope of this linear relationship. In this manner, we can see how the reaction process evolves.

During cure an epoxy resin changes from a viscous fluid to a rigid glassy solid, and the viscoelastic behaviour of the reacting mixture is expected to change as the molecular architecture varies during the reaction. The early stages of cure are often monitored by measurement of a shear viscosity [6–10]. An alternative to the study of the curing process is to use small amplitude dynamic measurement [11–16].

The reaction system may be considered as a gelation process. The evolution of the system between the sol phase and gel phase is a function of the extent of reaction. On the basis of percolation theory [17] as functions of the extent of reaction the zero rate viscosity is:

$$\eta_0 \approx |\alpha_c - \alpha|^{-k} \quad \text{if } \alpha_c \geq \alpha , \qquad (4)$$

while the zero shear rate modulus above the gel point is:

$$G_0 \approx |\alpha - \alpha_c|^\mu \quad \text{if } \alpha \geq \alpha_c . \qquad (5)$$

The percolation theory gives universal values for the exponents. From Eqs. (4) and (5) we see that the zero shear viscosity and the zero shear modulus are an unambiguous function of the extent of reaction. Thus, in Eq. (2) we can use the times belonging to the same viscosities at different temperatures below the gel point and the times belonging to the same zero shear moduli beyond the gel point.

In this study, we would like to investigate the kinetics of curing process by thermoanalytical and rheological method using some results of sol–gel transition theory.

Experimental work

Materials used in the investigation

The resin matrix consisted of a carboxyl terminated polyester resin crosslinked with triglycidyl isocyanurate. Full details are as follows:

- *Carboxyl terminated saturated polyester resin.*
  The resin used was a carboxyl terminated saturated polyester resin, manufactured from ethylene glycol, neopentyl glycol, trimellitic and phtalic anhydryde. It has the following properties:
  Acid value = 35
  Number average molecular mass = 1135
- *Triglycidyl isocyanurate.*
  Epoxy content = 9.3 value/kg
  Epoxy equivalent mass = 107 g/equivalent
  Molecular mass = 297 g/mol
  Melting point = 98 °C
  Density = 0.72 g/ml
- *Powder coating.*

This was a commercially available, electrostatically applied powder coating based on the above-mentioned resin matrix. The system contains 40% additives (fillers, pigments). The particle size of additives is 75 mm.

Sample preparation

The individual materials were first ground down to pass through a 100-micron sieve. They were subsequently mixed in the correct mixing ratio by grinding them together, using a mortar and pestle. The polyester resin to TGIC curing agent ratio was 93 parts of polyester resin to 7 parts of TGIC by mass.

Differential scanning calorimetry (DSC)

DSC analyses were carried out by means of a Perkin Elmer DSC-7 equipped with microprocessor. The instrument was calibrated by means of indium. The analyses were done in flowing nitrogen atmosphere (20 cm^3/min). Samples of material to be tested were prepared as described above, and weighed out to the nearest 0.01 mg into aluminium sample holders and sealed before analysis. A correction was applied to the power coating mass due to the fact that it contained only 60% of the polyester-TGIC resin matrix.

Isothermal DSC analysis

Isothermal DSC analyses were carried out on five different cure temperatures (160°, 180°, 200°, 220° and 240 °C). The curing time was 30 min. The dynamic analysis samples were heated from 50 to 300 °C at a heating rate of 10 °C/min.

From isothermal DSC measurements the conversion versus time curves were calculated.

Rheological methods

The complex viscosity and complex modulus of the systems were determined by a Rheometric Dynamic Stress Rheometer (RS-500) under oscillatory dynamic mode. A parallel plate measuring head was used with 25-mm plate diameter and a 0.2-mm gap. The measurements have been carried out with a small strain (5%) at five frequency values (1.9, 1, 0.7, 0.5, 0.3 rad/s) and at five temperatures (433, 443, 453, 463 and 473 K). In each case the components of complex modulus and the complex viscosity were continuously registered versus time.

From the measured complex viscosity values the zero shear viscosities were determined by extrapolation of the different frequency values to the zero frequency. At the beginning of measurements the viscosity of reaction mixture was determined. The viscosity of pigmented system was calculated from

$$\eta = \eta_m (1 + 2, 5\Phi)^b , \qquad (6)$$

where η_m is the viscosity of resin matrix, Φ is the volume fraction of filler, b is a constant. The value of b was determined by comparing the viscosities of pigmented and unpigmented mixtures. A similar equation was used in the case of zero shear modulus. The values of zero shear modulus were calculated from the measured storage modulus values by extrapolation to the zero frequency.

Results and discussion

From isothermal DSC measurements on the basis of conversion versus time curves the activation energy values were calculated for the different extent of reaction by Eq. (2). The activation energy was a constant value in a broad range of extent of reaction. In case of the resin matrix an activation energy value 64 kJ/mol was found between 0 and 0.7 extent of reaction, this value changed to 55 kJ/mol value beyond 0.7 extent of reaction and remained constant till 0.92 extent of reaction, then slowly decreased. In case of powder coating system the behaviour was similar, but the values of the activation energy were 54 kJ/mol and 51 kJ/mol, respectively.

From the viscosity data an activation energy value 63.2 kJ/mol and 56 kJ/mol were obtained in case of resin matrix and powder coating, respectively. These values are characteristic for the reaction carried out in the pregel state. From the zero shear modulus values an activation energy 52 kJ/mol and 48 kJ/mol were calculated for the resin matrix and the powder coating, respectively. These values characterise the reaction carried out in the postgel state.

It seemed from these preliminary calculations that there are three important processes in this curing system, one of them play role in the pregel state, and the other two in the postgel state.

If we investigate the first three reactions the change in concentration of hydroxyl terminated polymer can be represented by

$$\frac{d\alpha}{dt} = k_1(1 - \alpha)(R - \alpha) + k_3\alpha(1 - \alpha) - k_2\alpha(R - \alpha) , \qquad (7)$$

where α is the extent of reaction [$\alpha = (E_0 - E) - E_0$], $R = C_0/E_0$, E is concentration of epoxy group, C is the concentration of carboxyl group, E_0 and C_0 are the initial concentration values of epoxy and carboxyl groups, respectively, k_1, k_2 and k_3 are the reaction rates of reactions [1], [2] and [3], respectively.

Equation (7) can now be written as follows:

$$\frac{d\alpha}{dt} = a\alpha^2 + b\alpha + c , \qquad (8)$$

where the a, b and c are:

$a = k_1 - k_2 + k_3 ,$

$b = k_3 + Rk_2 - (R - 1)k_1 ,$

$c = Rk_1 .$

Equation (8) has an exact solution, but we used the differential form with a curve-fitting process.

In the early stage of reaction it is sufficient to take into consideration the first term of the right side of the Eq. (7). Thus the solution of the modified Eq. (7) is:

$$\ln\frac{1-\alpha}{R-\alpha} = -(R-1)k_1 t + \ln\frac{1}{R}. \tag{9}$$

For the second stage of reaction it can be supposed that the rate-determining term of Eq. (7) is the second or third terms of the right side, and thus the equation (7) has the following simplified solution:

$$\ln\frac{R-\alpha}{\alpha} = -Rk_i(t - t_*) + \ln\frac{R-\alpha_*}{\alpha_*}, \tag{10}$$

where $R = 1$ and $k_i = k_3$, if we take into account the second term of the right side of Eq. (7), and $R = R$ and $k_i = k_2$ in case of the validity of the third term in Eq. (7). α_* and t_* mean that extent of reaction and time, respectively from which the second or third term of the right side of Eq. (7) are the rate determining process.

In Table 1 the rate constants and activation energies calculated by means of Eqs. (8) and (9) are summarised. The results of two calculation methods agreed well in the region of extent of reaction between 0 and 0.7. On the basis of these results it seems probable that in the first stage of

curing process the first reaction is the rate-determining step, and this stage can be characterised by Eq. (9).

Table 2 shows the values of k_3 calculated by Eqs. (8) and (10). The curve fitting process gave a value of goodness acceptable (> 0.99) if we used $R = 1$ and $\alpha_* = 0.7$ and the region of extent of reaction was 0.7 ... 0.92.

In Table 3 the k_2 values calculated by the Eqs. (8) and (10) are given for the final stage of reaction ($\alpha = 0.92 ... 1$).

In both last cases the results calculated by the two methods agreed well.

On the basis of these calculations it seems to us that the curing process can be approximated by three simple relationships; in the first stage Eq. (9) can be used till 0.7 value of extent of reaction; in the second stage (between 0.7 and 0.92 extent of reaction) the Eq. (10) with $R = 1$ and $a_* = 0.7$, and at the final stage of curing process Eq. (10) may be used with $R = R$ and $a_* = 0.92$.

The zero shear viscosity is given by the general law:

$$\eta_0 = AM^a f(T - T_g), \tag{11}$$

where A is a constant, M is the molar mass of polymer, a is a constant, T_g is the glass transition temperature of polymer. Equation (11) is valid generally in the case of linear, monodisperse polymers, and the value a is 1 below a critical molecular mass and 3.4 beyond this critical value. In

Table 1 k_1 values calculated by Eqs. (8) and (9)

Temperature (°C)	Equation (8)		Equation (9)	
	Resin matrix (10^3 s^{-1})	Powder coating (10^3 s^{-1})	Resin matrix (10^3 s^{-1})	Powder coating (10^3 s^{-1})
160	1.55	0.77	1.42	0.92
180	3.49	2.12	3.33	1.96
200	8.28	2.75	7.89	3.02
220	13.4	2.07	12.8	2.00
240	13.5	1.47	13.0	1.30
Activation energy (kJ/mol)	65.0	54.2	66.4	51.0
$\ln k_0$	11.855	8.029	11.923	7.228

Table 2 k_3 values calculated by Eqs. (8) and (10)

Temperature (°C)	Equation (8)		Equation (10)	
	Resin matrix (10^3 s^{-1})	Powder coating (10^3 s^{-1})	Resin matrix (10^3 s^{-1})	Powder coating (10^3 s^{-1})
160	1.20	0.90	1.00	0.95
180	4.24	1.50	5.50	1.90
200	5.80	2.70	9.70	2.60
220	9.80	5.05	10.8	4.30
240	19.1	7.20	11.5	6.60
Activation energy (kJ/mol)	58.3	49.5	52.5	44.5
$\ln k_0$	9.664	6.7119	8.2386	5.4052

Table 3 k_2 values calculated by Eqs. (8) and (10)

Temperature (°C)	Equation (8)		Equation (10)	
	Resin matrix (10^5 s^{-1})	Powder coating (10^5 s^{-1})	Resin matrix (10^5 s^{-1})	Powder coating (10^5 s^{-1})
160	1.20	1.40	0.95	1.00
180	2.10	2.50	1.00	1.30
200	3.70	4.00	3.00	2.80
220	4.90	4.80	4.20	3.40
240	7.20	6.70	7.30	4.10
Activation energy (kJ/mol)	41.1	35.4	43.2	35.9
$\ln k_0$	0.1364	−1.2390	0.1578	−1.5910

Table 4 k_1 and k_3 values calculated from the rheological data

Temperature (°C)	Resin matrix		Powder coating	
	k_1 (10^3 s^{-1})	k_3 (10^3 s^{-1})	k_1 (10^3 s^{-1})	k_3 (10^3 s^{-1})
160	2.00	0.85	0.95	0.45
170	3.60	1.30	1.70	1.10
180	6.50	3.70	3.10	1.70
190	7.40	5.20	2.50	2.40
200	8.90	5.40	1.30	1.50
Activation energy (kJ/mol)	63.6	47.7	64.7	44.4
$\ln k_0$	11.5930	6.2031	11.764	4.7200

case of branched and polydisperse polymer the molar mass dependence of zero shear viscosity is more complex. There is a possibility to take into consideration this complex behaviour if we use some empirical a value in Eq. (11). The molecular mass of polymer relates to the extent of reaction:

$$M = \frac{M_0}{1 - \alpha}, \tag{12}$$

where M_0 is the molecular mass of the reaction components. Equations (11) and (12) give:

$$(1 - \alpha) = M_0 \left(\frac{A_1}{\eta_0}\right)^{1/a}, \tag{13}$$

where A_1 is a constant which contains the temperature dependence of zero shear viscosity. From Eqs. (9) and (13) and taking into account that $\ln(1 + x) \cong x$, we obtain a kinetic equation of the first phase of reaction:

$$\frac{\eta_0^{1/a}}{M_{0_1} A_1^{1/a}} = -\frac{R - 1}{2} k_1 t + \frac{1}{2} \ln \frac{1}{R}. \tag{14}$$

Thus, if we depict the zero shear viscosity versus t we can calculate the value k_1 from the slope and the interception

of straight line. The value of a was determined on the basis of curve-fitting process.

Table 4 shows the reaction rate value calculated by Eq. (14). The goodness of curve-fitting process was higher than 0.996 if we used value $a = 1$ below an extent of reaction 0.15 and $a = 3.4$ between the extent of reaction 0.15 and 0.7.

Beyond the extent of reaction 0.7 Eq. (10) was used to calculate the reaction rates and the values of extent of reaction were calculated from the zero shear modulus:

$$\alpha \cong \left(\frac{G_0}{G_\infty}\right)^{1/3}, \tag{15}$$

where G_∞ is the final value of the zero shear modulus.

Table 4 shows the k_3 values calculated by Eqs. (10) and (15). The values agreed well enough with the values calculated from the DSC measurements.

It should be noted that at higher temperatures some decreases can be seen in the rates of reactions which arise from the destruction of triglycidyl cyanurate. These values were not taken into account in the activation energy calculations.

Conclusions

The curing kinetics of unpigmented and pigmented carboxyl terminated polyester-triglycidylisocyanurate systems was investigated by thermoanalytical and rheological methods. On the basis of determination of activation energies versus extent of reaction we could divide the curing process into three characteristic stages. In the first stage, reaction [1] is the rate-determining process and this stage may be written by a simple equation (Eq. (9)).

At the second phase the rate-determining processes are the reactions [2] and [3] and this phase can be characterised by Eq. (10). The change in curing kinetics is carried out in the neighbourhood of the gel point. An approximating rheological calculation method was introduced and compared with the DSC method. This rheological evaluation method gave good enough results in the calculation of activation energy. The filler content generally decreases the rate of reaction in every case.

References

1. Schechter L, Wynstra L (1956) Ind Eng Chem 48:86
2. Schechter L, Wynstra L, Kurbjy RE (1956) Ind Eng Chem 48:94
3. Schechter L, Wynstra L, Kurbjy RE (1957) Ind Eng Chem 49:1107
4. Misev A (1991) Powder Coatings, Chemistry and Technology, Wiley and Sons, Chichester
5. Kamal MR, Sourour S, Ryan M (1973) SPE ANTECH Tech Papers 19:187
6. Tanaka Y, Kakiuchi H (1963) J Appl Polym Sci 7:1951
7. Kamal MA (1974) Polym Eng Sci 14:231
8. Roller MP (1970) Polym Eng Sci 15:406
9. Gonzales-Romero VM, Macosko CV (1985) J Rheol 29:1985
10. Hwang JG, Row CG, Lee SJ (1994) Ind Eng Chem 33:2377–2383
11. Tung CYM, Dynes PJ (1982) J Appl Polym Sci 27:569
12. Winter HA, Chambon J (1986) J Rheol 30:967
13. Winter HA (1987) Polym Eng Sci 27:1698
14. Cheng KC, Chin WY, Hsien H, Ma CC (1994) J Material Sci 29:283
15. Ganani E, Higgins BG, Powell RL (1986) Polym Eng Sci 26:1563
16. Plazek DJ, Choy IC (1991) J Appl Polym Sci Part B: Polym Phys 29:17
17. Stauffer D (1985) Introduction to percolation theory, Taylor and Francis, London

Progr Colloid Polym Sci (1996) 102:82–85
© Steinkopff Verlag 1996

Thermoreversible gelation in syndiotactic polystyrene/solvent systems

T. Roels
F. Deberdt
H. Berghmans

Dr. T. Roels · F. Deberdt
H. Berghmans (✉)
Laboratory for Polymer Research
Katholieke Universiteit Leuven
Celestijnenlaan 200F
3001 Heverlee, Belgium

Abstract Thermoreversible gelation in solutions of syndiotactic polystyrene (sPS) is investigated. Gelation in the solvents toluene, chlorobenzene, 1,2-dichlorobenzene, chloroform, 1,4-dioxane, tetrahydrofuran, o-xylene, cis- and trans-decalin is obtained by quenching the solutions in an ice bath.

Depending on the temperature and solvent-type, two gel types can be formed. Elastic, hazy gels (type I) can be obtained in all solvents by quenching in an ice bath. Cooling of the solutions in cis- and trans-decalin to high temperatures, and heating of type I gels, leads to the formation of past-like, opaque gels (type II). This gel type melts at higher temperatures than gel type I.

The solvent-dependence of this behaviour is related to the stability domains in the temperature-concentration diagram of different crystalline modifications. In cis- and trans-decalin, a transformation from the helical phase (gel type I) to the zigzag phase (gel type II) can take place on heating.

Key words Thermoreversible gelation – syndiotactic polystyrene – phase behaviour

Introduction

Thermoreversible gelation of polymer solutions is a well-known physicochemical process, frequently used by nature for the formation of solid-like systems that are composed of almost pure solvent. Typical examples are gelatin and carrageenans, and these polymers gelify in water at polymer concentrations below 1%. The mechanism of this gel formation is based on the intermolecular association of regular helices. Strong intra- and intermolecular forces like hydrogen bonding and ionic interactions are at the origin of this structure formation and the strong solvating power of water, generally used as the solvent, contributes in an important way.

The formation of similar gels with synthetic polymers can be found when synthetic polyelectrolytes are involved and the activities in this field are very important and deal with both physical and chemical gelation. The thermoreversible gelation of solutions of polymers wherein only Van der Waals interactions are involved (London dispersion forces or dipole interactions) is less evident. The most investigated polymer is syndiotactic poly(methyl methacrylate). This polymer cannot be crystallized from the melt but forms transparent gels in many solvents as there are toluene and o-xylene [1–3]. The gelation by specific interaction with its isotactic isomer was already mentioned in the paper that reported its synthesis [4].

This problem received renewed attention about 20 years ago when the first data on the gelation of solutions of isotactic polystyrene (iPS) were reported [5, 6]. This polymer crystallizes from the melt with a 3_1 helix conformation. The chains adopt a different conformation in the almost transparent gels that are obtained in, for example

Progr Colloid Polym Sci (1996) 102:82–85
© Steinkopff Verlag 1996

decalin. Most of the experimental data suggest that gelation proceeds by the formation of a 12_1 helix [7, 8] although modificaitions of the 3_1 helix have also been proposed [9, 10].

When syndiotactic polystyrene (sPS) came available, it rapidly became clear that this behaviour was not unique for iPS. The syndiotactic isomer is also capable of forming two different conformations. The all trans, T_4 conformation is found in melt crystallized samples, and samples crystallized from solution under specific conditions. Crystallization can also proceed from solution with the formation of a T_2G_2 helix conformation [11–13]. Recent investigations of the phase behaviour of sPS in different solvents have shown that crystallization with the T_4 conformation corresponds to the well known folded chain lamellar crystallization. The crystalline phase that is based on the T_2G_2 helix formation represents an incongruently melting polymer-solvent compound [14–16]. The formation of two types of gels can therefore be expected and the purpose of this paper is to bring additional information on this topic.

Experimental

Syndiotactic polystyrene was supplied by Dow Chemical, USA. The number-average and mass-average molecular masses, determined by GPC in 1,2,4-trichlorobenzene at 135 °C, are 14.3×10^4 and 42.9×10^4.

Solutions with a concentration of 5% by weight were prepared in glass tubes and sealed. These gels were used for optical inspection.

Thermal behaviour was studied by Dynamic Scanning Calorimetry (DSC2 and 7). Crystalline phases were identified by wide-angle x-ray scattering patterns recorded with a Philips PW 1792 flat film camera, using CuK_α radiation. More details can be found in previous publications [14–16].

Gel formation

Solutions of the as-received sPS can only be prepared by heating the polymer and the solvent to high temperature for a long period of time. Homogeneous solutions have been obtained in the following solvents: toluene, chlorobenzene, 1,2-dichlorobenzene, chloroform, 1,4-dioxane, tetrahydrofuran, o-xylene, cis- and trans-decalin.

Under these experimental conditions, it was not possible to make solutions in acetone and diethyl ether.

When the homogeneous solutions are brought back to lower temperature by quenching in an ice bath, rigid, transparent gels are obtained that turn slightly hazy on standing. This type of gel will be called type I gels.

Cooling the solutions in cis-decalin and trans-decalin to temperatures above 130 °C leads to the formation of paste-like, very opaque gels. We call these gels type II gels. This structure is maintained when the samples are further cooled to room temperature.

Gel melting

The melting of these gels is followed by the test tube tilting method. For this purpose, the glass tube is placed upside down in the oil bath and the temperature raised at 2 °C/min. The onset of flow is taken as the gel melting temperature.

Two different gel melting processes have been observed and can be related to the type of solvent.

Melting of the gel in toluene, o-xylene, THF, chloroform, 1,4-dioxane, chlorobenzene and 1,2-dichlorobenzene

In these solvents, the gel (type I) remains unchanged up to the final melting temperature. A transparent solution is obtained.

The observed gel melting temperatures are listed in Table 1. Melting temperatures between 120° and 130 °C are found in toluene, o-xylene and the chlorobenzenes. Lower melting points are found in chloroform (72 °C), 1,4-dioxane (95 °C) and tetrahydrofuran (85 °C). These last three melting points are situated above the normal boiling point of the solvent.

These high gel melting points explain the difficulties encountered when preparing solutions of sPS, especially in low boiling solvents. In these solvents, solutions can only be prepared at pressures higher than atmospheric pressure, in a sealed environment.

Table 1 Melting temperatures of the gels, compared with boiling points of the solvent

	T_{melt} (°C)	T_{boil} (°C)
chloroform	72	61
tetrahydrofuran	75	67
toluene	118	111
o-xylene	120	144
1,4-dioxane	95	101
chlorobenzene	130	132
1,2-dichlorobenzene	127	179
cis-decalin	150	193
trans-decalin	150	185

Melting of the gel in cis- and trans-decalin

Preparation in the way described above results in gels that
are very similar to the previous ones: elastic and hazy.
When gels of type I prepared in these two solvents are
heated, no change is observed on heating unless the temperature is raised above 125 °C. An important change in
transparency takes place. The gel turns from hazy to
opaque. During this change, the gel remains solid-like. It is
clear that during heating, a transformation from type
I into type II takes place. On further heating, a change in
the optical aspect of the opaque gel sets in at 150 °C. The
opalescence decreases and flow sets in before complete
disappearance of the opalescence. A further increase of
the temperature completes the gel to sol transition and
a transparent solution at high temperature is obtained.

Discussion

The melting behaviour is observed to be highly solvent-
quality dependent. In the solvents cis- and trans-decalin,
two gel structures with different properties can be prepared. In the other solvents, only one gel structure is
formed.

The gel formation and melting processes can be well
understood in the framework of the temperature-concentration diagrams published earlier [14–16]. In these diagrams, melting points obtained from DSC-measurements
are plotted versus concentration. The temperature at the
end of the observed endotherm in DSC is reported as the
melting point. These melting temperatures correspond
very well with the optically observed ones.

In Fig. 1, the phase diagram of sPS in chlorobenzene is
shown. It can be used as a model for the other solvents in
which only one type of gelation (type I) is observed. The
phase behaviour of sPS in these solvents has been described in detail elsewhere [14–16]. Gels can be prepared
between 1% and 20% polymer concentration. This concentration region is indicated in the phase diagram by the
shaded area. Concentrations lower than 1% do not form
a coherent gel structure, unless a very high molecular mass
is used. Samples with concentrations exceeding 20% are
difficult to prepare.

Study of the phase behaviour revealed that in this
domain the δ-phase, in which the polymer chain backbone
adopts a helical conformation, is formed. This helical
phase represents an incongruent melting polymer-solvent
compound. The high degree of transparency is characteristic for the absence of large lamellar crystals that are
generally formed by crystallization from the melt. The
morphology seems to be more of the bundle-like type as
proposed for the very similar gelation of the isotactic

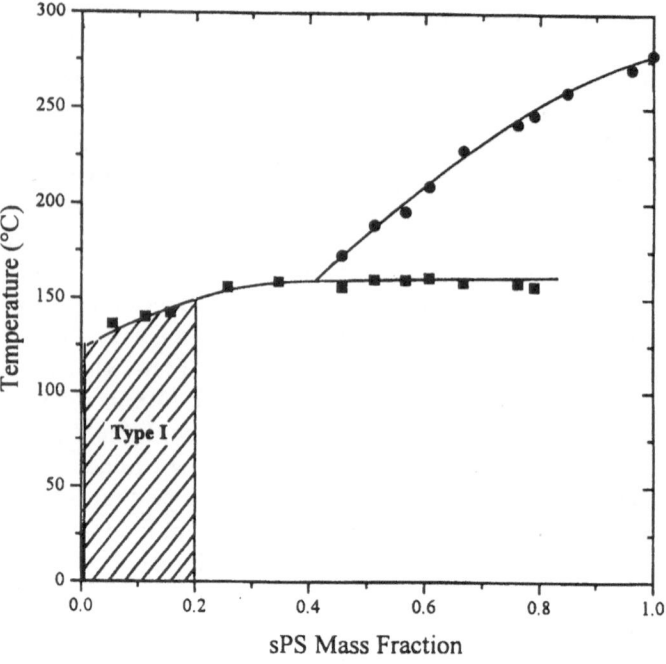

Fig. 1 Temperature-concentration diagram of the system sPS/
chlorobenzene: ■: melting of the helical phase; ●: melting of the
zigzag phase. The shaded area represents the domain in which gel
type I can exist

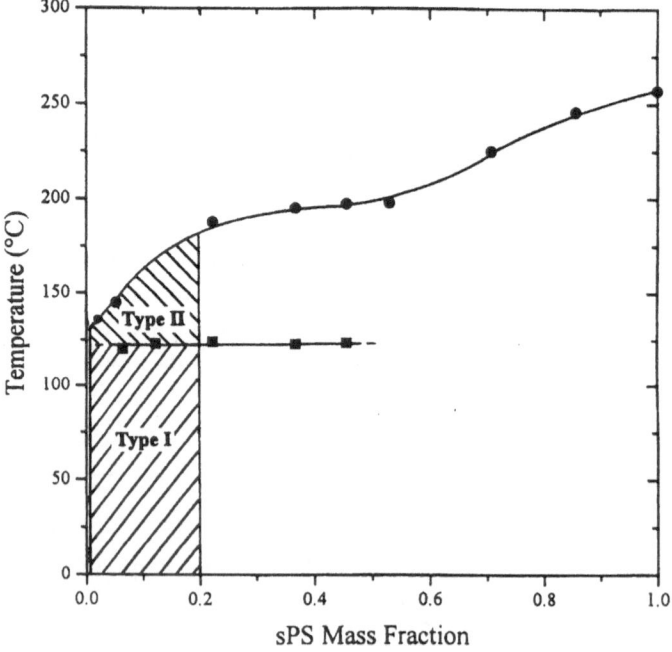

Fig. 2 Temperature-concentration diagram of the system sPS/cis-
decalin: ■: melting of the helical phase; ●: melting of the zigzag phase.
The shaded areas represent the domains in which different types of
gel can exist

Progr Colloid Polym Sci (1996) 102:82–85
© Steinkopff Verlag 1996

isomer [17]. This type of gelation is also observed in solutions of biopolymers like gelatin and carrageenan.

The temperature-concentration diagram of the system sPS/cis-decalin is represented in Fig. 2. A similar behaviour is observed in the trans isomer. The region of interest for gelation is indicated on the diagram. In this region, two melting lines are present. The lower one represents the melting of the helical phase. This delimits the stability domain of the hazy gel-structure (type I). The same gelation mechanism as the one observed in chlorobenzene is responsible for this gelation.

Increasing temperature induces melting of the helical phase and recrystallization into the β-phase. The polymer changes its conformation from a helical structure to a zigzag structure. This transition is very fast, so that the macroscopic properties of the gel are not lost and that the system can maintain its solid-like macroscopic structure. The observed opalescence of the type II gels is attributed to the presence of lamellar folded crystals in which the polymer backbone adopts a zigzag conformation. The upper melting line represents the melting of these crystals, and the corresponding gel structure. At even higher temperature, the system is a homogeneous solution.

From these observations we can conclude that two types of gel can be formed in cis- and trans-decalin: a hazy gel at low temperature (type I), and an opalescent, paste-like one at higher temperatures (type II). In both structures, the molecular conformation differs and can be related to the stability domains of the corresponding phases.

References

1. Berghmans H, Donkers A, Frenay L, Stoks W, De Schrijver FC, Moldenaers P, Mewis J (1986) Polymer 28:97–102
2. Berghmans M, Thys S, Cornette M, Berghmans H, De Schrijver FC, Moldenaers P, Mewis J (1994) Macromolecules 27:7669–7676
3. Spevacek J, Schneider (1987) Adv in Colloid and Interface Sc 27:81–150
4. Fox FG, Garret BS, Goode WE, Gratch S, Rincaid JF, Spell A, Stroupe JD (1958) J Am Chem Soc 80:1768–1769
5. Jones D, Latham A, Keller A, Miyasaka (1973) J Pol Sc Pol Phys Ed 11:1759–1767
6. Atkins E, Isaak D, Keller A, Miyasaka K (1977) J Pol Sc Pol Phys Ed 15:211–226
7. Corradini P, Guerra G, Petraccone V, Pirozzi B (1980) Eur Polym J 5:1089–1092
8. Sundararajan P, Tyrer N, Bluhm T (1982) Macromolecules 15:286–290
9. Guenet JM (1986) Macromolecules 19:1960–1968
10. Chatani Y, Nakamura N (1993) Polymer 34:1644–1648
11. Grassi A, Longo P, Guerra G (1989) Makromol Chem Rapid Commun 10:687–690
12. Kobayashi M, Nakaoki T, Ishihara N (1989) Macromolecules 22:4377–4382
13. Guerra G, Vitagliano V, De Rosa C, Petraccone V, Corradini P (1990) Macromolecules 23:1539–1544
14. Deberdt F, Berghmans H (1993) Polymer 34, 10:2192–2201
15. Deberdt F, Berghmans H (1994) Polymer 35, 8:1694–1704
16. Roels T, Deberdt F, Berghmans H (1994) Macromolecules 27:6216–6220
17. Atkins EDT, Hill MJ, Jarvis DA, Keller A, Sarhene E, Shapiro JS (1984) Colloid Polym Sci 262:22–45

Progr Colloid Polym Sci (1996) 102:86–88
© Steinkopff Verlag 1996

M.Ye. Solovjev
V.A. Kapranov
V.I. Irzhak
A.G. Galushko

The mechanical properties of elastomers with chemical and physical junctions

Dr. M.Ye. Solovjev (✉)
Maksimova b.7, app. 3
150000 Yaroslavl, Russia

M.Ye. Solovjev · V.A. Kapranov
A.G. Galushko
Yaroslavl State Technical University
Moscow avenue 88
150053 Yaroslavl, Russia

V.I. Irzhak
Institute of Chemical Physics
Chernogolovka
Moscow distr., 142432, Russia

Abstract Theoretical model of the mechanical properties of elastomer network with thermoreversible junctions have been obtained on the basis of non-equilibrium thermodynamics and physical kinetics methods. Non-Gaussian assumption for a free energy of chains in the equilibrium state have been used. As was established, the possibility of construction of the tearing master-curve is determined by the energy of junctions. If the energy of junctions is less then 60 kJ/mol the time-temperature superposition principle contravenes and anomalous large fracture work is observed. It was found that ultimate tensile strength versus junctions energy curve has a maximum. Thus, an optimal value of junctions energy exists.

Key words Elastomers – networks – physical junctions – mechanical properties

As is known, there are many anomalies in the fracture and the mechanical properties of elastomer networks with thermoreversible junctions [1–3]. In our previous studies [4], the theoretical model of the mechanical properties of such system was obtained on the basis of non-equilibrium thermodynamics and physical kinetics methods. It was assumed that deformations are not too large and the Gaussian assumption for a free energy of chains in the equilibrium state is acceptable. The appropriate experiments with nonylacrylate-acrylamide copolymers have been provided.

In the present work, we consider the case when the strain value of the elastomer network with thermoreversible junctions is sufficiently large and, therefore, not only reversible process of breakdown of junctions, but also irreversible process of chains degradation can arise (fracture of the network).

The expressions for non-equilibrium mechanical stress of the network can be obtained similar to [4]:

$$\sigma_{ne} = -T\frac{dS}{d\lambda} + \eta \tag{1}$$

$$\frac{d\eta}{dt} = -\eta \bigg/ \left(\frac{r}{j} + \frac{1}{j}\frac{d\sigma_e}{d\lambda}\right), \tag{2}$$

where S and σ_e are the entropy and equilibrium mechanical stress correspondingly (these parameters have been calculated from partition function of the network), r and j have determined from molecular parameters of the system, λ is the strain value. For numeric calculations expression σ_e has been represented as expansion by a strain as follows

$$\sigma_e = vkT\bigg\{\left(1 - \frac{1}{N}\right)\left(\lambda - \frac{1}{\lambda^2}\right) + \frac{3}{5N}\left(\lambda^3 - \frac{4}{3\lambda^3}\right)$$

$$+ \frac{1}{5N^2}\left[22\left(\lambda - \frac{1}{\lambda^2}\right) - \frac{159}{5}\left(\lambda^3 - \frac{4}{3\lambda^3}\right)\right.$$

$$+ \frac{93}{7}\left(\lambda^5 + \frac{3}{5}\lambda^2 - \frac{8}{5\lambda^4}\right)$$

$$+ \frac{1}{25N^3}\left[-\frac{253517}{1120}\lambda + \frac{322253}{1120\lambda^2} + \frac{8166}{5}\left(\lambda^3 - \frac{4}{3\lambda^3}\right)\right.$$

$$- \frac{10923}{7}\left(\lambda^5 + \frac{3}{5}\lambda^2 - \frac{8}{5\lambda^4}\right)$$

$$+ 537\left(\lambda^7 + \lambda^4 - \frac{64}{35\lambda^5}\right)\bigg]\bigg\} + \cdots, \tag{3}$$

Progr Colloid Polym Sci (1996) 102:86–88
© Steinkopff Verlag 1996

Fig. 1 Tensile strength of the network as function of the ultimate strain for different energy of junctions (E)

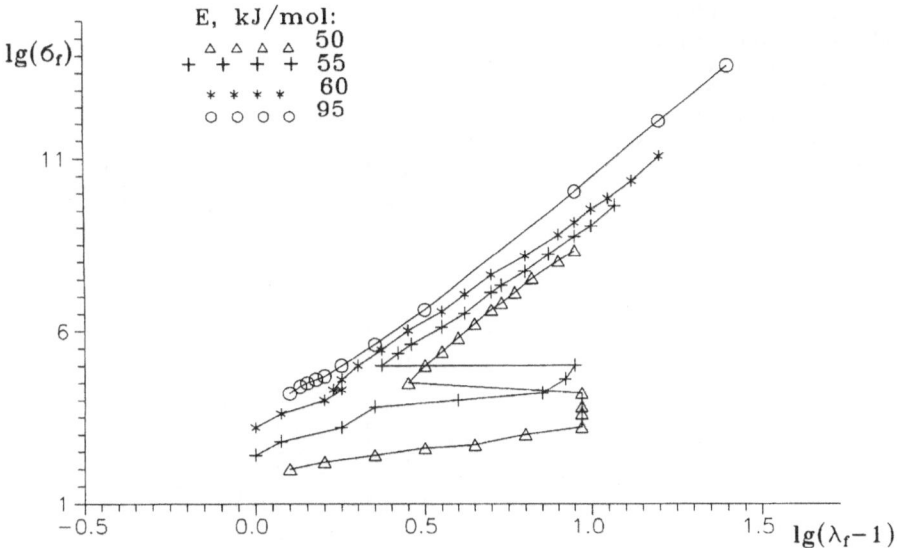

where v is the chain concentration, N is the number of segments in a chain. These equations have been added with kinetic equations of physical junctions concentration,

$$\frac{dg}{dt} = \frac{G}{\sqrt{N}}\left\{ -\exp\left[\lambda'^2 + \frac{2}{\lambda'} - 3 - \frac{E - TS}{kT}\right]g^{3/2} \right.$$

$$\left. + (n - g)^{3/2}\right\} \qquad (4)$$

$$\sqrt{\lambda_u^2 + \frac{2}{\lambda_u} - 3} = \sqrt{\lambda^2 + \frac{2}{\lambda} - 3}\left(1 - \frac{\sigma}{2gkT}\frac{1}{\lambda - \frac{1}{\lambda^2}}\right), \qquad (5)$$

$$\lambda' = \frac{\lambda}{\lambda_u} \qquad (6)$$

where g is the junctions' concentration, G is the constant, E and S are the energy and entropy of junctions.

For description of the mechanical fracture, we use the kinetic chain destruction equation

$$\frac{dv}{dt} = \frac{vNkT}{h}\exp\left\{ -\frac{E_0}{kT} + \frac{S_N \neq S_N}{k}\right\}, \qquad (7)$$

where k and h are the Boltzmann and Plank constants, T is the absolute temperature, E_0 is the chain destruction energy, and $S_N \neq S_N$ is the difference between chain entropies in a basic state and activated state.

In this model tensile strength of the network is the stress value when chain concentration became equal to zero. We have considered the viscoelastic properties and tensile strength depending on the thermodynamical para-

meters of physical junctions: energy, entropy and maximal concentration determined by chemical structure of chains. In Fig. 1 tensile strength of the network as function of the ultimate strain curves for different energy of physical junctions are presented. For many kinds of amorphous elastomers such curves are considered as master-curve of fractures as was investigated by Smith [5]. As we see in the Fig. 1, the possibility of master-curve construction is obtained only if the energy of junctions is similarly high as chemical ones. If the energy of junctions became less than 60 kJ/mol, we can see the anomalous large fracture strain that leads to increase of fracturing. The reason for this phenomena is capability of the physical junctions to break reversibly. As a result, we see that such gels do not obey WLF time-temperature superposition principle.

Therefore, we conclude that capabilities of physical junctions, to unequilibrium energy dispersion leads to increasing of network fracture stability. In this relation it is interesting to investigate the tensile strength and ultimate strain of the network in dependence of the junctions energy. These curves are presented in Fig. 2.

Maximum on the tensile strength curve substantiates that there is an optimal value of the junctions energy. This result may be interesting for chemists attempting to synthesize new polymers with high mechanical properties. For the practical application of the reversible junction concept in elastomer network, we have obtained the simple model of filled elastomer. As supposed, physical junctions are created on the polymer-filler interface border. Reversible fracture of these bonds on the filler surface may be a factor that is responsible for the reinforcement of composition by filler.

M.Ye. Solovjev et al.
The mechanical properties of elastomers

Fig. 2 Tensile strength (σ_f)
and ultimate strain (λ_f) in
dependence on junctions'
energy (E). Concentration
of functional groups that
determines maximal junctions'
concentration is: $1'$, $2'$–10%,
1, 2–15%

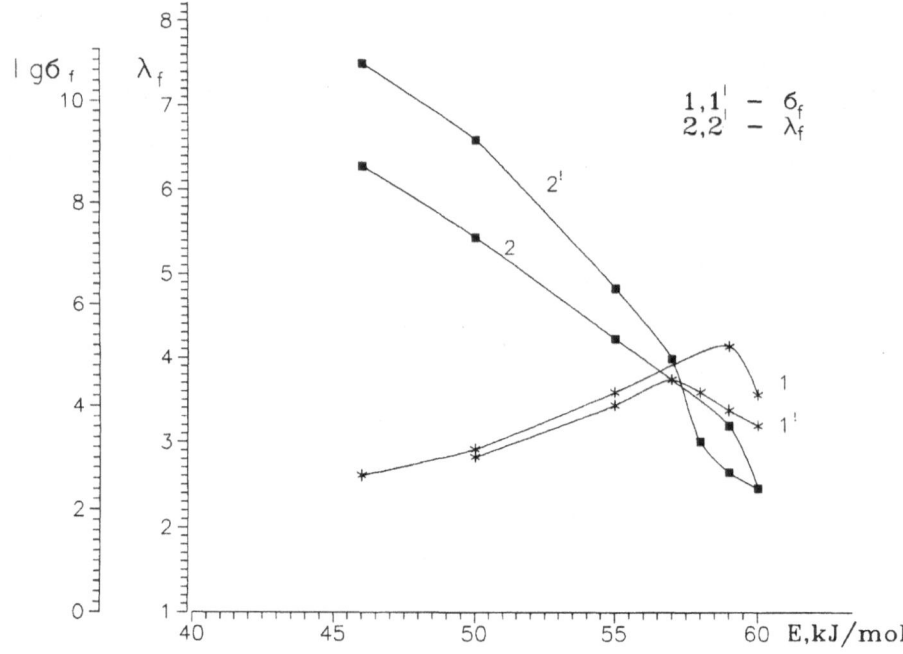

References

1. Halpin JC (1965) Rubber Chem and Technol 38:1007–1037
2. Hamed GR (1983) Rubber Chem and Technol 56:244–251
3. Hnat V, Proborova E, Raab M (1991) Macromol Chem Macromol Symp 41: 247–251
4. Solovjev MYe, Raukhvarger AB, Basaev AR, Privalov AN, Irzhak VI, Korolev GV, Makhonina LI (1992) Polymer Sci, USSR 34:172–174
5. Smith TL (1958) J Polymer Sci 32:99

Progr Colloid Polym Sci (1996) 102:89–97
© Steinkopff Verlag 1996

Ö. Pekcan
Y. Yılmaz

Fluorescence method to study gelation swelling and drying processes in gels formed by solution free radical copolymerization

Dr. Ö. Pekcan (✉) · Y. Yılmaz
Department of Physics
Istanbul Technical University
Maslak, 80626, Istanbul, Turkey

Abstract Gelation during solution free radical crosslinking copolymerization (FCC) of methyl methacrylate (MMA) and ethylene glycol dimethacrylate (EGDM) was studied by using the steady state fluorescence technique. FCC was performed in the presence of toluene at 75 °C. Sol–gel transitions were monitored by observing the direct intensity of an excited pyrene (P_y) during in situ fluorescence experiments for various toluene content. Gel points were determined and critical β exponents were found to be around 0.38. In situ swelling and drying experiments were performed at 50 °C by monitoring P_y intensity in disk-shaped gels produced by solution FCC. Li–Tanaka equation was employed to produce the cooperative diffusion coefficient, D_c for swelling process. D_c values were found to be around 10^{-6} cm^2/s at 50 °C. It was also observed that loosly formed gels swell much faster than densely formed gels.

Key words Fluorescence – percolation – gelation – swelling – drying

Introduction

In polymeric systems a gel is a cross-linked polymer network swollen in a monomeric liquid medium where the liquid prevents the polymer network from collapsing into a compact mass. The polymer network of a gel can be formed in various ways [1]. In condensation polymerization, a network is formed by polymerizing bifunctional units and polyfunctional units. The bifunctional units form linear chains and polyfunctional units serve as cross-linkers. A polymer network can also be formed by cross-linking polymers formed from bifunctional monomers by free radical polymerization. Gelation process occurs during a random linking process of subunits to larger and larger molecules. The critical point, called gel point, divides the liquid and the solid phases. This critical point always exists if the system is disordered and if all processes are random [2]. At gel point the system behaves neither as a liquid nor as a solid or any time and length scale. Scaling theory provides a basis for modeling sol–gel phase transition. Various models for this liquid–solid transition have been proposed; the most well known are percolation theory [3] and an aggregation approach [4]. The percolation model can predict critical exponents for gel fraction, weight average degree of polymerization, radius of gyration, etc., near the sol–gel phase transition. The critical exponents in percolation theory differ from those found in classical theories of Flory and Stockmayer [1, 5].

Polymer networks or gels are known to exist generally in two forms, swollen and shrunken. Volume transitions may occur between these forms either continuously or sudden jumps between them [6, 7]. The equilibrium swelling and shrinking of gels in solvents have been extensively studied [8–10]. The swelling, shrinking and drying kinetics of physical and chemical gels are very important in many technological applications, especially in pharmaceutical industries in designing slow-release devices for oral drugs. In understanding cosmetic ingredients the mechanism of swelling, shrinking and drying kinetics is highly desirable.

In the agricultural industry, for producing storable foods, and in medical applications in developing artificial organs the knowledge of gel kinetics is an important requirement.

The swelling properties of chemically cross-linked gels can be understood by considering the osmotic pressure versus the restraining force [11–15]. The total free energy of a chemical gel consists of bulk and shear energies. In fact, in a swollen gel the bulk energy can be characterized by the osmotic bulk modulus K, which is defined in terms of the swelling pressure and the volume fraction of polymer at a given temperature. On the other hand, the shear energy which keeps the gel in shape can be characterized by shear modulus G. Here, shear energy minimizes the nonizotropic deformations in gel. The theory of kinetics of swelling for a spherical chemical gel was first developed by Tanaka et al. [16] where the assumption is made that the shear modulus, G, is negligible compared to the osmotic bulk modulus. Later, Peters et al. [17] derived a model for the kinetics of swelling in spherical and cylindrical gels by assuming non-negligible shear modulus. Recently, Li and Tanaka [11] developed a model where the shear modulus plays an important role which keeps the gel in shape due to coupling of any change in different directions. This model predicts that the geometry of gel is an important factor and swelling is not a pure diffusion process. Several experimental techniques have been employed to study the kinetics of swelling, shrinking and drying of chemical and physical gels among which are neutron scattering [18], quasielastic light-scattering [17], macroscopic experiments [12] and in situ interferrometric [19] measurements.

In this work, we carry out in situ fluorescence experiments to study gelation, swelling and drying kinetics of chemically cross-linked gels. It is known that steady state fluorescence spectra of many chromophores are sensitive to their environment. The interactions between the chromophore and the solvent molecules affect the energy difference between the ground and the excited states. This energy difference is called the Stokes shift, and depends on the refractive index and dielectric constant of the solvent. Recently, by measuring the Stokes shift of a polarity sensitive fluorescent species, the gelation during epoxy curing was monitored as a function of cure time [20]. Time-resolved and steady-state fluorescence techniques were employed to study isotactic polystyrene in its gel state [21]. Pyrene derivative was used as a fluorescence molecule to monitor the polymerization, aging and drying of aluminosilicate gels [22]. Recently, we reported in situ observations of the sol–gel phase transition in free-radical crosslinking copolymerization, using the steady state fluorescence (SSF) technique [23, 24]. In this paper gelation, swelling and drying processes of gels formed by free radical

crosslinking copolymerization (FCC) of methyl methacrylate (MMA) and ethylene glycol dimethacrylate were studied using the quenching properties of the excited state of a fluorescing molecule. Pyrene (P_y) is used as a fluorescence probe to monitor gelation, swelling and drying processes during in situ fluorescence experiments.

Theoretical considerations

Percolation model

In lattice percolation model [25], monomers are thought to occupy the sites of a periodic lattice. Between two nearest neighbors of these lattice sites a bond is formed randomly with the probability p. Thus, for $p = 0$, no bonds have been formed and all monomers remain isolated clusters. However, in the other extreme, i.e., for $p = 1$ all monomers in the lattice have clustered into one infinite network. This network is called a gel and a collection of finite clusters is called a sol. Usually, there is a sharp phase transition at some critical point $p = p_c$, where an infinite cluster starts to appear. This point is called the gel point, for p below p_c only a sol exists, but for p above p_c, both sol and gel coexist together. Thus, gelation is a phase transition from a state without gel to a state with gel. The sol–gel transition occurs asymptotically near the sol–gel transition point; the gel fraction satisfies the following relation [25].

$$G = B(p - p_c)^\beta \tag{1}$$

with a suitable constant β, called critical exponent. The asymptotic proportionality factor B is referred to as the critical amplitude. During gelation, if solvent molecules are present in the system, then one should allow the sites to be occupied by a monomer with a probability ϕ (mol fraction of the monomers in the solvent) and to be occupied by a solvent molecule otherwise, with probability $(1 - \phi)$. Now two nearest-neighbor monomers may form a bound with probability p, however no bonds can be formed between solvent–solvent and solvent–monomer molecules. In this case random site bond percolation model has to be employed instead of random bond percolation model. In site bond percolation model, clusters connected by random bonds are formed from randomly distributed monomers.

Model for swelling and shrinking processes

It has been suggested [11] that the kinetics of swelling and shrinking of a polymer network or gel should obey the following relation

$$\frac{W(t)}{W_\infty} = 1 - \sum_{n=1}^{\infty} B_n e^{-t/\tau_n}, \tag{2}$$

where $W(t)$ and W_∞ are the swelling or solvent uptake at time t and at infinite equilibrium respectively. Here, B_n represents a constant related to the ratio of shear modulus G and longitudinal osmatic modulus, M which is defined by the combination of shear and osmotic bulk moduli as [14, 15] $M = 4/3G + K$. τ_n is the swelling rate constant. In the limit of large t or if τ_1 is much larger than the rest of τ_n, all high-order terms ($n \geq 2$) in Eq. (2) can be neglected, then Eq. (2) becomes

$$\frac{W(t)}{W_\infty} = 1 - B_1 e^{t/\tau_1}. \tag{3}$$

Here, B_1 is given by the following relation [11].

$$B_1 = \frac{2(3 - 4R)}{\alpha_1^2 - (4R - 1)(3 - 4R)}, \tag{4}$$

where $R = G/M$ and α_1 is a function of R, i.e.,

$$R = \frac{1}{4}\left[1 + \frac{\alpha_1 J_0(\alpha_1)}{J_1(\alpha_1)}\right]. \tag{5}$$

In Eq. (3), τ_1 is related to the collective cooperative diffusion coefficient D_c of a gel disk at the surface and given by the relation [12]

$$\tau_1 = \frac{3a^2}{D_c \alpha_1^2}. \tag{6}$$

Here, a represents half of the disk thickness in the final infinite equilibrium state which can be experimentally determined. The B_1–R and α_1–R relations are presented in ref. [11] for a disk-shaped gel.

Fluorescence method

Fluorescence and phosphorescence intensities of aromatic molecules are affected by both radiative and non-radiative processes [26]. If the possibility of perturbation due to oxygen is excluded, the radiative probabilities are found to be relatively independent of environment and even of molecular species. Environmental effects on non-radiative transitions which are primarily intramolecular in nature are believed to arise from breakdown of the Born–Oppenheimer approximation [27]. The role of the solvent in such a picture is to add the quasi-continuum of states needed to satisfy energy resonance conditions. The solvent acts as an energy sink for rapid vibrational relaxation which occurs

after the rate-limiting transition from the initial state. Years ago, Birks et al. studied the influence of solvent viscosity on fluorescence characteristics of pyrene solutions in various solvents and observed that the rate of monomer internal quenching is affected by solvent quality [28]. As the temperature of liquid solution is varied, the environment about the molecule changes and much of the change in absorption spectra and fluorescence yields in solution can be related to the changes in solvent viscosity. A matrix that changes little with temperature will enable one to study molecular properties themselves without changing environmental influence. Poly(methyl methacrylate) (PMMA) has been used as such a matrix in many studies [29]. Recently, we have reported viscosity effects on low frequency, intramolecular vibrational energies of excited naphthalene in swollen PMMA latex particles [30]. In this work these properties of aromatic molecules were used to monitor, first the sol–gel phase transition in FCC, and then the swelling and drying processes of this gel in a good solvent.

Experiments

Three types of in situ fluorescence experiments are planned to be carried out. At first, the sol-gel transitions in solution FCC of MMA and EGDM are probed by using the steady-state fluorescence technique. The radical copolymerization of MMA and EGDM was performed in toluene solution at 75 °C in the presence of 2,2′-azobis-isobutyronitrile (AIBN) as an initiator. P_y was used as a fluorescence probe to detect the gelation process, where below the "gel point" MMA, linear and branched PMMA chains act as an energy sink for the excited P_y, but above gel point the PMMA network provides an ideal unchanged environment for the excited P_y molecules. Naturally, from these experiments, one may expect a drastic increase in fluorescence intensity, I of P_y around the gel point [23, 24]. The gels obtained in these experiments were dried under vacuum and used in the experiments described below.

In the second and third types of in situ experiments, swelling and drying processes at 50 °C are probed by P_y molecules in gels obtained from the first type of experiments. In these experiments continuous transitions are expected which should result in continuous decrease or increase in I during swelling or drying processes respectively. Here, one may expect that as gel swells, solvent molecules act as an energy sink for the excited P_y molecules and I intensity should decrease due to quenching. However, as gel shrinks, drying network starts to provide an ideal, unchanged environment for the excited P_y molecules; as a result, I intensity is expected to increase.

EGDM has been commonly used as crosslinker in the synthesis of polymeric networks [31]. Here, for our use, the monomers MMA (Merck) and EGDM (Merck) were freed from the inhibitor by shaking with a 10% aqueous KOH solution, washing with water and drying over sodium sulfate. They were then distilled under reduced pressure over copper chloride. The polymerization solvent, toluene (Merck), was distilled twice over sodium.

In situ steady-state fluorescence measurements were carried out using a Perkin Elmer Model LS-50 Spectrometer equipped with temperature controller. All measurements were made at the 90 °C position and slit widths were kept at 2.5 mm.

Results and discussion

Gelation

In the first type of experiments, AIBN (0.26 wt%) was dissolved in MMA and this stock solution as divided and transferred into round glass tubes of 15 mm internal diameter for fluorescence measurements. Six different samples were prepared using this stock solution with various toluene contents for solution polymerization. Details of the samples are listed in Table 1. All samples were deoxygenated by bubbling nitrogen for 10 min and then radical copolymerization of MMA and EGDM was performed at 75 ± 2 °C in the spectrometer fluorescence accessory. Here, EGDM content was kept as 0.01 Vol% and P_y concentration was taken as 4×10^{-4} M. P_y molecule was excited at 345 nm during in situ experiments and variation in fluorescence emission intensity, I, was monitored at 395 nm with the time-drive mode of the spectrometer. No shift was observed in the wavelength of the maximum intensity of P_y and all samples have kept their transparency during the polymerization process. Normalized P_y intensities versus reaction times are plotted in Fig. 1 for samples with various toluene contents. Gelation curves in Fig. 1 represent asymptotic behaviours, which give evidence of typical critical phenomenon and may suggest that percolation theory can be employed to interpret these results.

In order to quantify the above results, we assumed that the reaction time, t for the polymerization is proportional to the probability p and the fluorescence intensity, I monitors the growing gel fraction G; then, Eq. (1) can be written as

$$I = A(t - t_c)^\beta . \tag{7}$$

Here, the critical time t_c corresponds to the gel point p_c, and A is the new critical amplitude. t_c can be determined by taking the first derivative of the experimentally

Table 1 Gelation parameters of toluene polymerization. All measurements were performed at 75 ± 2 °C. AIBN and EGDM contents were kept as 0.26 wt% and 10^{-2} vol% for all samples respectively

Gels	1	2	3	4	5	6
Toluene vol%	0.10	0.13	0.15	0.20	0.23	0.30
β	0.40	0.43	0.27	0.47	0.32	–
t_c (sec)	1715	1772	1885	2035	2950	2963

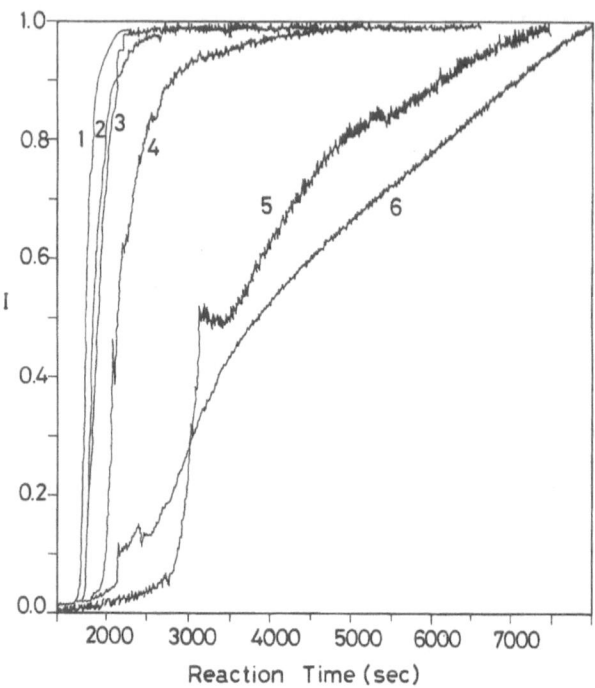

Fig. 1 Variation in P_y intensity I versus reaction time, t during toluene FCC. Numbers 1, 2, 3, 4, 5 and 6 correspond to samples in Table 1

obtained I curve with respect to t. The maximum in dI/dt curve corresponds to the inflection point in curve I, which gives the t_c, in time axis [24]. Below t_c, since P_y molecules are free, they can interact and be quenched by sol molecules; as a result, I presents small values. However, above t_c, since most of the P_y molecules are frozen in the network, I intensity gives very large values. The plots of $\log I = \log A + \beta \log(t - t_c)$, for the data shown in Fig. 1 produced β values, in the region $10^{-2} < |1 - t/t_c| < 10^{-1}$ above t_c. The critical exponents, β are listed in Table 1 together with t_c values for the gels produced in solution polymerization. Gelation times t_c are shifted to larger values as toluene content increases. However, average β value was found to be around 0.37, independent of the toluene content [24].

Progr Colloid Polym Sci (1996) 102:89–97
© Steinkopff Verlag 1996

Swelling and drying

In the swelling and drying experiments disk-shaped gels were used which were cut from the cylindrical gels obtained from the first type of experiments. Disk gels were placed on the wall of the square fluorescence cell (1 cm × 1 cm), so that the exciting light beam can only see the gel. After pouring toluene into the cell, P_y was excited at 345 nm and the variation in I was monitored at 395 nm, during in situ swelling experiments at 50 °C. No shift was observed in the maximum intensity of P_y before and after the swelling process is completed and all samples have kept their transparency during the swelling process. Figure 2 presents the steady state spectra of P_y before and after the swelling process is completed. Normalized P_y intensities versus swelling time are plotted in Fig. 3 for the gels prepared by toluene polymerization. The numbers in Fig. 3 correspond to gel samples listed in Table 1. It is interesting to note that solvent uptake is much faster in loosely formed gels (5 and 6) than densely formed gels (1 and 2).

After decanting toluene from the fluorescence cell, swollen gel, open to the air, was left to dry and the variation in I intensity was monitored at 395 nm during the in situ drying experiment at 50 °C. The behaviour of P_y intensities versus drying time are shown in Fig. 4 for the gel samples listed in Table 1. Here, all samples were excited at 345 nm and monitored at 395 nm during in situ

Fig. 2 Fluorescence emission spectra of P_y, before and after the swelling process. P_y molecules are excited at 345 nm

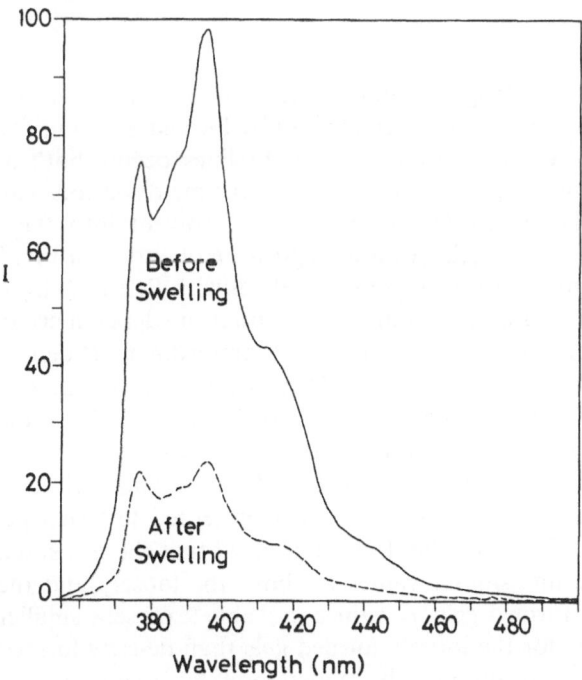

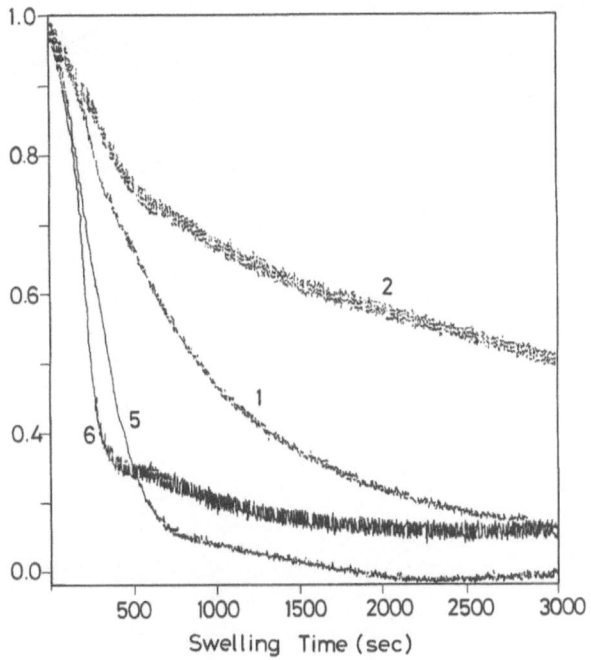

Fig. 3 P_y intensity I versus swelling time for the gels listed in Table 1

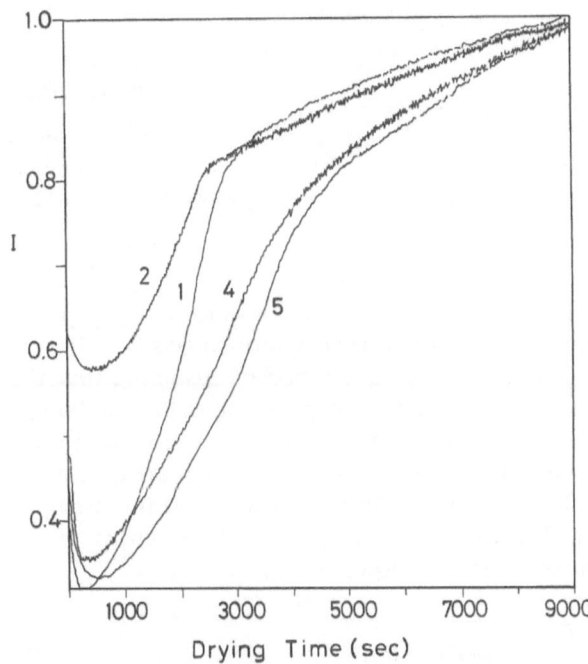

Fig. 4 P_y intensity I versus drying time for the gels given in Table 1

fluorescence experiments. If one compares the swelling and drying data shown in Figs. 3 and 4, one can reach a conclusion that swelling and drying processes are not symmetrical and, in general, gels swell much faster than they dry.

Fig. 5 Plot of the digitized data of Fig. 3, which obey the relation $\ln(1 - I_\infty/I(t)) = \ln B_1 - t/\tau_1$. a, b, c and d present the data of gels with numbers 1, 2, 5 and 6 in Table 1

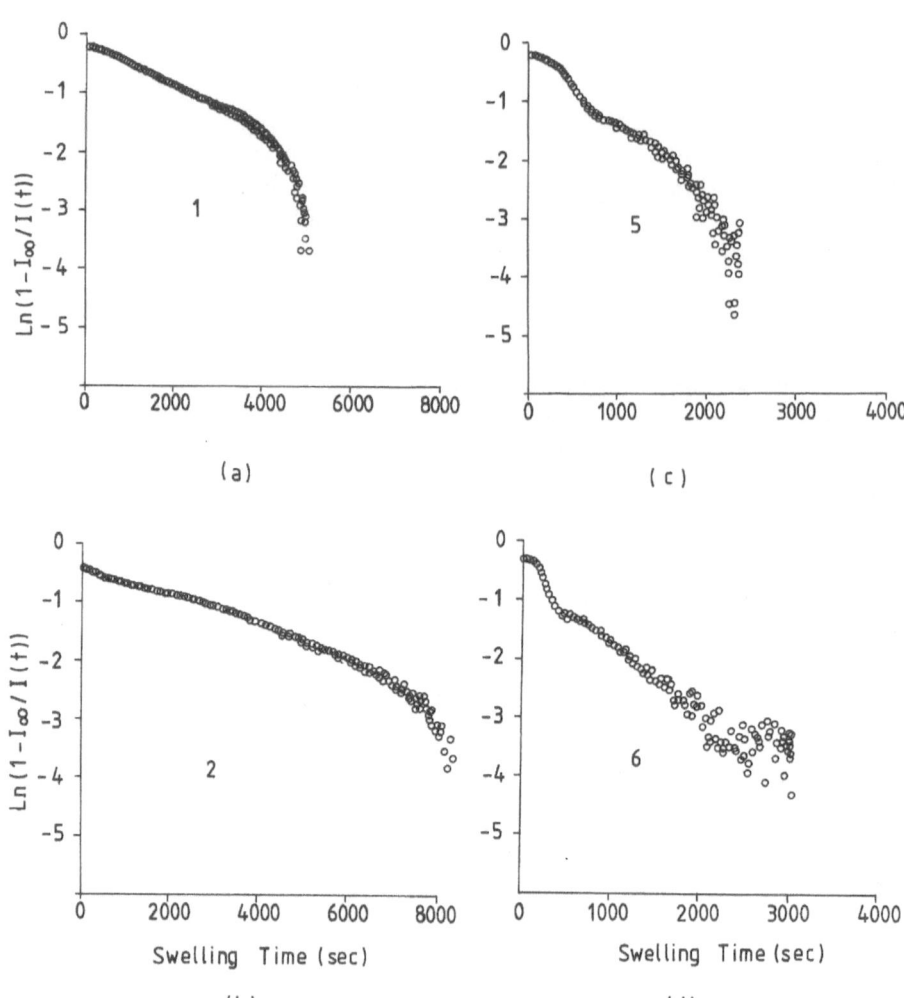

Here, it has to be also noted that loosely formed gels (4 and 5) dry much slower than densely formed gels (1 and 2).

These results can be quantified by assuming that the solvent uptake, W is inversely proportional to the fluorescence intensity I. During swelling, as more solvent molecules enter into the gel, more P_y molecules are quenched; as a result I decreases as W increases. Under this assumption, Eq. (3) can be written in terms of I intensity in the logarithmic form, as follows

$$\ln\left(1 - \frac{I_\infty}{I(t)}\right) = \ln B_1 - t/\tau_1 . \qquad (8)$$

The above arguments can hold for the drying processes in a reverse fashion, where during drying, as W decreases, I increases.

Figures 5a, b, c and d present the digitized data of Fig. 3 in the form of Eq. (8). For the densely formed gels (Figs. 5a and b), deviation from the straight lines occurs only at very long time limit where solvent uptake reaches an equilibrium. However, for loosely formed gels (Figs. 5c and d) deviation from the straight lines occurs both at short and long time limits. The short time deviations can be explained by fast penetration of solvent into these loosely formed gels. A linear regression of curves in Fig. 5 at intermediate time region provide us with B_1 and τ_1 from Eq. (8) (see Fig. 6). Taking into account the dependence of B_1 on R (ref. [11]), one obtains R, and from α_1–R dependence, α_1 values are found. Then, using Eq. (6), one can determine the cooperative diffusion coefficient, D_c of the disk-shaped gels. Experimentally obtained parameters, τ_1 and B_1 together with α_1 and D_c values are summarized in Table 2 where a values are also presented for each gel sample. Here, one should have noticed that the measured D_c, presents similar values for both the loosely and the densely formed gels. τ_1 values in Table 2 present smaller numbers for the loosely formed gels than densely formed gels. This result may be understood by assuming that

Fig. 6 Linear regression of the data presented in Figs. 5a, b, c and d by using Eq. (8). τ_1 and B_1 values were obtained from the slopes and the intersections of the linear curves, respectively

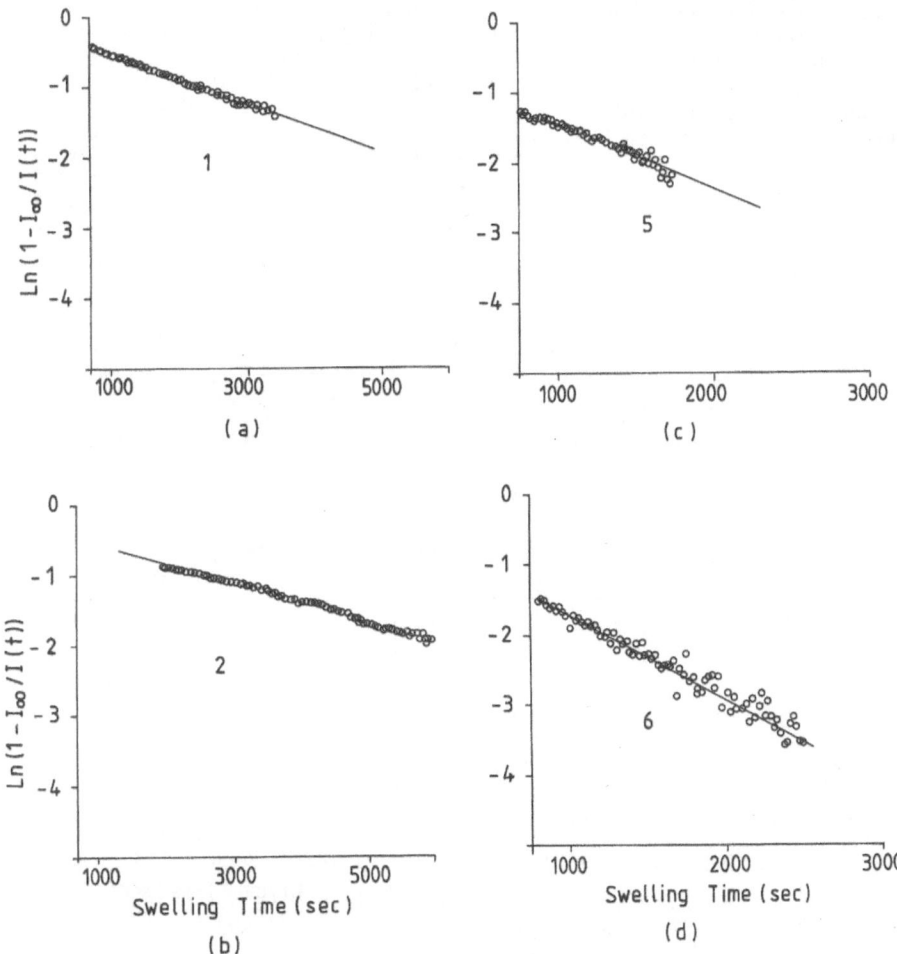

Table 2 Swelling parameters of disk-shaped gels given in Table 1. τ_1 and B_1 values were obtained from Eq. (9) by linear regression of the swelling data

Gels	1	2	3	4	5	6
τ_1 (sec)	2849	3571	1597	844	1067	810
B_1	0.85	0.78	0.38	0.49	0.61	0.62
α_1	1.39	1.66	2.3	2.3	2.12	2.09
$D_c \times 10^{-6}$ (cm^2/sec)	8.0	7.4	5.7	8.7	8.3	11.7
a (cm)	0.12	0.15	0.12	0.11	0.11	0.11

loosely formed gels have more vacant spaces, which can swell easily and take less time to reach their fully swollen state.

The drying data in Fig. 4 are digitized and plotted in Fig. 7, by obeying Eq. (8). The structure of these curves are all similar and have three distinct regions. At long time limit they all reached to saturation in fully drying state. At intermediate times curves in Fig. 7 behave linearly. At the short time limit, loosely formed gels (Fig. 7c and d) present more linear relation than densely formed gels (Fig. 7a and b). Even though the mechanism of drying process is quite complicated and much different than the deswelling process, just to have some feeling, we used Eq. (8) to interpret the drying data. B_1 and τ_1 values are produced by fitting the drying data to Eq. (8) at short and intermediate time regions. Produced B_1, τ_1, α_1 and D_c values are summarized in Table 3 for the gels given in Table 1. Here, it is interesting to note that average τ_1 values present larger numbers for the loosely formed gels than densely formed gels. This behaviour is opposite to the swelling process, which is understandable, because if gel takes more solvent it needs longer time to release it and to reach its full drying state. However, densely formed gels dry with the same rate as they swell. For the drying processes, we observed some meaningless B_1 values. At this stage of our work it is difficult to speculate on these results. D_c values are also found at least two times smaller in drying processes than in swelling processes. This issue is

Fig. 7 Plot of the digitized data of Fig. 4, which obey the relation $\ln(1 - I_{\infty}/I(t)) = \ln B_1 - t/\tau_1$. a, b, c and d present the data of gels with numbers 1, 2, 4 and 5 in Table 1

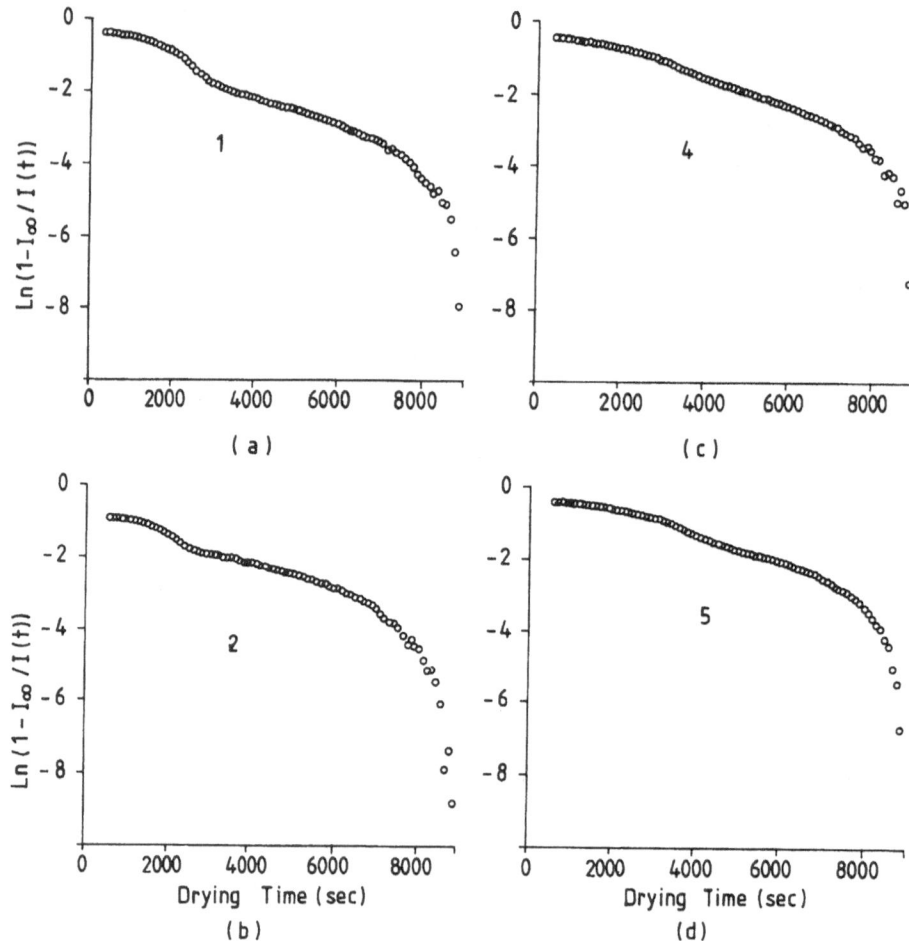

Table 3 Drying parameters of disk-shaped gels given in Table 1. τ_1 and B_1 values are obtained from Eq. (9) by linear regression of the drying data. Fits are done at both short and intermediate time regions

Gels	1	2	3	4	5	6
τ_1 (sec)	2710	1515	3759	5076	5000	5291
	2924	3598	1161	2427	2941	2525
B_1	0.88	–	0.68	0.71	0.80	0.63
	0.44	0.34	–	–	–	0.67
α_1	1.24	–	1.92	1.84	1.57	2.03
	2.3	2.3	–	–	–	1.97
$D_c \times 10^{-6}$ (cm²/sec)	10.6	–	3.4	2.3	3.2	1.9
	2.8	3.8	–	–	–	4.2
a (cm)	0.12	0.15	0.12	0.11	0.11	0.11

also highly speculative for our knowledge and open to discussion.

In conclusion, these preliminary results have shown that the direct fluorescence method can be used for real time monitoring of gelation, swelling and drying processes. In this novel technique in situ SSF experiment is easy to perform and we believe non stop monitoring of above processes are highly desirable in the field. At this stage of our work it is difficult to say more about the method; we need more experiments with different gel systems.

Acknowledgement We thank Professor O. Okay for supplying us with the material and his ideas.

References

1. Flory PJ (1953) Principles of Polymer Chemistry Cornell University, Ithaca NY
2. Family F, Landau DP (eds) (1984) Kinetics of Aggregation and Gelation North Holland, Amsterdam
3. Stauffer D (1985) Introduction to Percolation Theory, Taylor and Francis, London
4. Hermann H (1986) J Phys Rev 153:136
5. Sockmayer WH (1943) J Chem Phys 45:11
6. Dusek K, Paterson D (1968) J Polym Sci A2 6:1209
7. Tanaka T (1980) Phys Rev Lett 45:1636

Progr Colloid Polym Sci (1996) 102:89–97
© Steinkopff Verlag 1996

8. Tobolsky AV, Goobel JC (1970) Macromolecules 3:556
9. Schild HG (1992) Prog Polym Sci 17:163
10. Amiya T, Tanaka T (1987) Macromolecules 20:1162
11. Li Y, Tanaka T (1990) J Chem Phys 92(2):1365
12. Zrinyi M, Rosta J, Horkay F (1993) Macromolecules, 26:3097
13. Candau S, Baltide J, Delsanti M (1982) Adv Polym Sci 7:44
14. Geissler E, Hecht AM (1980) Macromolecules 13:1276
15. Zrinyi M, Horkay F (1982) J Polym Sci Polym Ed 20:815
16. Tanaka T, Filmore D (1979) J Chem Phys 20:1214
17. Peters A, Candau SJ (1988) Macromolecules 21:2278
18. Bastide J, DuOplessix R, Picot C, Candau S (1984) Macromolecules 17:83
19. Chi Wu, Chni-Ying Yan (1994) Macromolecules 27:4516
20. Lin KF, Wang FW (1994) Polymer 4:687
21. Wandelt B, Birch DJS, Imhof RE, Holmes AS, Pethnick RA (1991) Macromolecules 24:5141
22. Panxviel JC, Dunn B, Zink JJ (1989) J Phys Chem 93:2134
23. Pekcan Ö, Yılmaz Y, Okay O (1994) Chem Phys Lett 229:537
24. Pekcan Ö, Yılmaz Y, Okay O, Polymer (in press)
25. Stauffer D, Coniglio A, Adam M (1982) "Gelation and Critical Phenomena" Advances in Polym Sci Springer-Verlag, Berlin, Heidelberg 74:44
26. Kropp LJ, Dawson RW (1969) "Fluorescence and Phosphorescence of Aromatic Hydrocarbons in Polymethylmethacrylate" Molecular Luminescence on international Conference, Ed. Lim, EC, New York, Benjamin
27. Bixon M, Jortner J (1968) J Chem Phys 48:715
28. Birks JB, Lumb MD, Munro IH (1964) Proc Roy Soc A 277:289
29. Jones PF, Siegel J (1969) J Chem Phys 50:1134
30. Pekcan Ö (1995) J Apply Polym Sci 57:125
31. Okay O, Gürün Ç (1992) J Apply Polym Sci 46:421

Progr Colloid Polym Sci (1996) 102:98–100
© Steinkopff Verlag 1996

A ^{13}C-NMR study of polyacrylic acid gels as radioactive ion sorbents

A. Saidel Vasilescu
C.C. Ponta

Dr. A. Saidel Vasilescu (✉) · C.C. Ponta
Institute of Physics and Nuclear Engineering
S.1, NMR Laboratory
P.O. Box MG-6
Bucharest, Magurele, Romania

Abstract A ^{13}C-NMR study was performed on polyacrylic acid (PAA) gel swollen in water in the presence of some di- and trivalent ions (Ca^{2+}, Sr^{2+}, Ba^{2+}, Zn^{2+}, UO_2^{2+}, Al^{3+}). As concentration of metal ions grows, the NMR signals (CH_2, CH, COO) gradually broaden up to the spectrum disappearance. This is the moment when the gel collapses and the metal ion saturation occurs. The information obtained from ^{13}C-NMR can be used in radioactive waste treatment procedures.

Key words ^{13}C-NMR – polyacrylic acid – gels applications – radioactive waste treatment

Introduction

The known property of the polyanionic gels to retain multivalent ions [1] opened a new application field of these gels, as sequestering agents in radioactive waste treatment.

One of the typical polyanions is the polyacrylic acid (PAA). A crosslinked PAA is obtained by gamma induced polymerization of the acrylic acid. Various structures could be designed in a simple manner. The swelling degree, the most important macroscopic property of these hydrogel networks, are connected with the crosslinking density. The higher swelling degree, the more porous is the network structure and the more easy it is for an ion to defuse into this structure.

This work studies by high resolution ^{13}C-NMR spectroscopy the effect produced on NMR signals of the PAA hydrogel due to the presence of di- and trivalent ions, some of them having radioactive correspondents. The aim of this study is to observe the modifications of the PAA constitutive chain groups, induced by the increase of the metal ion concentration; moreover, we wish to answer the question of whether the ^{13}C-NMR spectroscopy is able to control this absorption process.

Our experiments in the radioactive waste treatment showed a good decontaminate capacity of the PAA hydrogel.

Experimental

A crosslinked PAA is obtained in our institute by gamma induced polymerization of the acrylic acid. The used PAA had a swelling degree of 40 g/g in pure water. Its preparation was described elsewhere [2]. Superabsorbent hydrogels, with a swelling degree more than 100 g/g, are useless for NMR measurements because the concentration of polymer in the swollen gel is too small.

Inorganic salts of di- and trivalent ions used were of analytical purity: $CaCl_2$, $Sr(NO_3)_2$, $BaCl_2$, $UO_2(NO_3)_2$, $ZnCl_2$, $Al(NO_3)_3$.

Stock solutions in water (3.3 M) of each of these salts were prepared and gradually added (aliquots of 0.1 ml) to 0.454 g polymer, in powder form, expanded into 2 ml of water.

^{13}C-NMR measurements were performed with a Bruker FT-SXP 90 equipment, operating at 22.63 MHz for ^{13}C resonance, with proton broad band decoupling, internal deuterium lock on capillary tubes filled with C_6D_6, room temperature, SW = 6000 Hz, number of scans 2000–4500, repetition time 2 s, LB = 5.

The measurements were performed 30 min after salt solution addition to the hydrogel under stirring. Each sample was recorded several times, in order to check the achieving of equilibrium; averaged signal widths were taken into consideration.

Results and discussion

The polyacrylic acid $(-CHCOO^- -CH_2)_{nn}H^+$ has in the swollen state the negative electric charges along the macromolecular chain and the counter ions (H^+) more or less free.

In the swollen state three forces (pressures) are acting: *the elastic force* (rubber like), which tends to contract/expand the gel, in function of the equilibrium lengths of the polymer strands; *the polymer strands affinity*, depending on the electric properties of the molecules and finally, *the hydrogen-ion pressure*, determined by the hydrogen ions existing in the gel fluid as a result of the ionization of the polymer network.

The ^{13}C-NMR spectrum of the PAA hydrogel consists of three broad signals (Fig. 1a) for the three constitutive groups, i.e., COO^-, CH and CH_2. Their chemical shifts are, respectively: 184; 45.2 and 36.1 ppm (relative to TMS as internal standard). The obtained broad signals (the

averaged widths at room temperature is 25 Hz for carboxylic group and 40 Hz for methylenic and methinic groups) suggest that segmental motion of polymer chains is sufficiently fast to give interpretable NMR spectra. The fact that the COO^- signal is narrower signifies a greater mobility of this hydrophilic group by comparison with the hydrophobic groups CH and CH_2.

The addition of metal ions caused a continuous broadening of the hydrogel PAA ^{13}C-NMR signals. This effect is notable in the CH and CH_2 case, where the broadening is very sensitive to the metal ion concentration. The carboxylic signal is less influenced.

A further increase of the ion concentration leads to the gel collapse. The available negative charges along the polymer chain are neutralized by the metal ions and saturation occurs. When saturation takes place, the collapse of the PAA gel is visible. The water is expelled from the gel as a result of a syneresis process. This is manifested in the ^{13}C-NMR spectrum by the disappearance of the signals (Figs. 1b, c). This phenomenon takes place for all studied metal ions, but for different saturation degrees. We define

Table 1 ^{13}C-NMR signal width variation of PAA gel with ion saturation

Salt	Sat. degr. %	COO^-(Hz)	CH(Hz)	CH_2(Hz)	Ionic radius (Å)
$CaCl_2$	0	25	29.4	50	0.99
	10	25	41.15	52	
	20	26	115	85	
	30	40	–	–	
	35	–	–	–	
$Sr(NO_3)_2$	0	25	30	42	1.13
	10	28	34	45	
	20	33	97	163	
	30	54.5	100	212	
	40	74	265	250	
	50	98	–	–	
	55	–	–	–	
$BaCl_2$	0	25	30	42	1.35
	20	27	45.6	67	
	40	28	72	79.5	
	50	50	91	100	
	60	85	130	230	
	70	100	–	–	
	80	–	–	–	
$ZnCl_2$	0	25	36	45	0.74
	10	30	49	70	
	20	85	–	–	
	25	–	–	–	
$UO_2(NO_3)_2$	0	26	36	45	≤ 0.85
	10	30	56	53	
	20	33	68	58	
	30	–	–	–	
$Al(NO_3)_3$	0	25	30	48	0.50
	10	90	100	235	
	15	–	–	–	

Fig. 1 NMR spectra of PAA gel: a) swollen in water; b) in the presence of metal ions

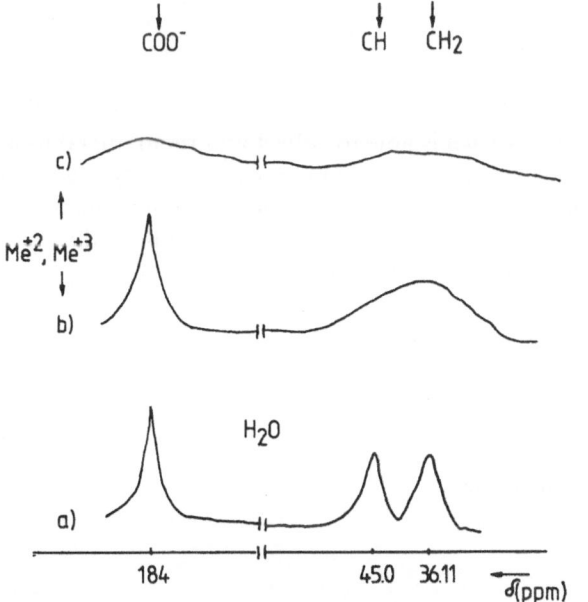

the saturation degree as the ratio between the molar concentration of cation and the available carboxylic groups concentration. In Table 1 are given the values of the NMR signal widths (Hz) at different saturation degrees with several ions.

The metal salt addition neutralizes the electric charges within the network and has as result the reducing of the hydrogen ion pressure on one side, and the increasing of the polymer strand affinity, on the other side, both of them contributing in the same sense, i.e., to the gel shrinkage.

One may observe from Table 1 that, under similar conditions, the saturation degree at which the phase separation is produced depends on the electric charge carried by ion (more precisely on the ionic radius) and on the ion valence. Thus, for the studied ions, the saturation degree that produces the gel collapse varies directly proportional to the ionic radius and inversely proportional to the metal valence.

Practically, the ion concentration necessary to saturate a similar quantity of PAA hydrogel decreases in the order: $Ba^{2+} > Sr^{2+} > Ca^{2+} > UO_2^{2+} > Zn^{2+} > Al^{3+}$.

The monovalent ions serve to substitute the role of the solvent molecules, from the point of view of the network dimensions; they are attracted by the carboxylic groups, displacing the solvent polar molecules, without generating new cross links. The multivalent ions have a different effect: they are located near the negative charged groups, moving away from the water molecules and gathering two or more carboxylic groups, whether the metal is di or trivalent. This creates new crosslinks that shorten the distances between the polymer strands; when all the available carboxylic groups are implied in the new crosslinks, the saturation occurs and the collapse takes place. According to [3, 4], the PAA gel in the presence of monovalent cations behaves as a H type and in the multivalent cation presence, as an L type.

The disappearance of the NMR signals signifies the loss of the segmental mobility of the PAA polymer. The signal corresponding to carboxylic groups is the last which "vanished". We presume that these groups still maintain more mobility than the chain groups, being a sort of flexible cross joint. In the metal ion presence the rest of the network becomes more rigid, due to the increased polymer strands affinity, and the movement of the carboxylic groups reduces up to spectral evanescence. This spectral behavior prompted us to consider that ^{13}C-NMR spectroscopy can be a good monitor of the ion retention process.

The sensitive retention of the metal ions by PAA hydrogels leads to the idea of using them as sequestering agents for contaminant ions. Indeed, the decontaminate power of PAA hydrogels was manifested in our experiments performed with the radioactive liquids containing ions as: $^{85+89}$Sr, ^{60}Co, ^{137}Cs, which were treated with PAA gels. The simplicity and the efficiency of the procedure recommended these gels in waste treatment technology. Table 2 summarizes the retention capacity of PAA gel used on different surfaces contaminated with $^{85+89}$Sr. The efficiency of decontamination is independent on contaminated substrate, as can be observed. The polyacrylate salts, shrunk hydrogels, contain very little quantity of water and can be treated in radioactive waste technology as solid residues. A recovery procedure is also possible, PAA acting as an ion-exchanger. If this one is preferred, the recovery control can be performed also by ^{13}C-NMR spectroscopy, the only technique able to detect the emergence of "free" PAA signals.

Table 2 The use of PAA gel in $^{85+89}$Sr decontamination of different materials

Material	A_0 (counts/s)	A_f(counts/s)
aluminum	200	20
stainless steel	266	6
PVC	340	10

where A_0, A_f-initial and final rate counting (counts/s).

Conclusions

- The PAA gel is able to collect and retain metal ions;
- The efficiency of the sequestering property is greater with smaller ionic radius of the contaminant element;
- ^{13}C-NMR spectroscopy is able to follow this process, as well as the gel recovery.

References

1. Skouri R, Schosseler F, Munch J, Candau SJ (1995) Macromolecules 28:197–210
2. Ponta CC, Seitan G, Tudusciuc G (1992) Proc Intl Biomed Eng Days Istanbul: 224–226
3. Axelos M, Mestdagh M, Francois J (1994) Macromolecules 27:6594–6602
4. Ikegami A, Imai N (1962) J Polym Sci 56: 133–149

Progr Colloid Polym Sci (1996) 102:101–109
© Steinkopff Verlag 1996

L. Hegedüs
M. Wittmann
N. Kirschner
Z. Noszticzius

Reaction, diffusion, electric conduction and determination of fixed ions in a hydrogel

L. Hegedüs · M. Wittmann · N. Kirschner
Prof. Z. Noszticzius (✉)
Department of Chemical Physics
Technical University of Budapest
1521 Budapest, Hungary

Abstract Connecting the aqueous solutions of a strong acid and a strong base with a hydrogel cylinder in electric field creates a special non-equilibrium chemical system called "electrolyte diode", where reaction, diffusion and electric conduction can be studied simultaneously. It was found that the current-voltage characteristic of such a diode is strongly affected by relatively small concentrations of fixed ions in the hydrogel. A theory to explain the role of the fixed ions is proposed. According to that theory, from the slope and the intercept of the current-voltage characteristic curve the fixed ion concentration can be easily calculated. Experiments with pure and polyelectrolyte modified PVA based hydrogels support the proposed theory.

Key words Hydrogels – fixed ions – polyelectrolytes – electric conduction – current-voltage characteristic

Introduction

Chemical reaction–diffusion patterns are in the focus of current interest. When these patterns are generated in free liquid systems convection of the liquid is a major experimental problem. A natural choice to avoid convection is to apply a gel medium allowing reaction and diffusion but preventing convection. Following this straightforward idea different chemical reaction–diffusion patterns like chemical waves [1–3], Turing structures [4–6] or self-replicating chemical spots [7] were studied in open reactors applying hydrogel media. In these experiments the only intended role of the gel was to suppress convection. Nevertheless, in some cases [8, 9] it was found that the gel played a more active role and the gel itself [8] or some of its components [4, 9, 10] participated in the pattern-forming reaction. It was also found that the gel can affect the Liesengang ring structures [11], a classical pattern forming reaction–diffusion phenomenon.

These observations suggest that in a far from equilibrium situation, when strong chemical potential gradients are present in the gel, some new effects can appear which could be characteristic not only for the reaction–diffusion system but for the gel medium as well. Recently, we studied a system [12] where beside chemical concentration gradients an electric potential gradient can be established and maintained in a hydrogel reactor. Here, we report that in this complex and non-linear system the current-voltage characteristic is determined not only by the mobile and chemically reacting and diffusing species, but it is also strongly affected by minute concentrations of fixed ionic groups in the hydrogel.

In our special system pH and electric potential differences are applied simultaneously on a polyvinylalcohol –glutardialdehyde hydrogel cylinder. The schematic drawing of the most important parts of the apparatus is shown in Fig. 1. A theoretical current-voltage characteristic – or polarization – curve of such a system is presented in Fig. 3. As the characteristic curve in Fig. 3 is similar to that of a semiconductor diode the above system is referred to as "electrolyte diode" [12–14]. The theoretical polarization curve of such an electrolyte diode was calculated based on the Nernst–Planck equations and the electroneutrality condition, but assuming an "ideal" gel without any fixed charge in the gel. Obviously, when fixed ionic groups are present in the gel the appropriate equations should be involved in this theory. Before doing that, however, in this introductory part we summarize briefly the main results of the electrolyte diode theory without fixed ions [12, 13]. The extended theory will be presented in the Appendix of this paper.

To understand the behavior of an electrolyte diode, let us regard a gel cylinder connecting aqueous solutions of a strong base and a strong acid (see Fig. 1). For example, let us assume a KOH solution of concentration c_0 on one side and a HCl solution with the same c_0 concentration on the other side of the gel. In "forward" direction the electric field is applied so that K^+ ions from the KOH and Cl^- ions from the HCl solutions migrate into the gel where they form a KCl solution of c_0 concentration. For large forward voltages most of the gel is filled with that solution. Thus, the absolute value of the current density $|i|$ in this case is given by the following formula:

$$|i| = F \cdot (D_K + D_{Cl}) \cdot c_0 \cdot \frac{|\Delta \varphi|}{l}, \qquad (1)$$

where F is the Faraday number, D_K and D_{Cl} are the ionic diffusion coefficients, $|\Delta \varphi|$ is the absolute value of the dimensionless voltage: $|\Delta \varphi| = (F/RT)|U|$, and l is the effective length of the gel cylinder connecting the alkaline and acidic solutions. (Absolute values $|i|$ and $|\Delta \varphi|$ are used in the formulae throughout this paper instead of i and $\Delta \varphi$ to avoid misunderstanding or unnecessary complications due to different sign conventions. The sign convention used in the figures regards the forward current as positive and the backward current as negative.) U is the voltage drop in the gel in Volts. In the forward direction the current is carried mainly by K^+ and Cl^- ions and the contribution of H^+ and OH^- ions is negligible.

The situation in the so-called "backward" direction is qualitatively different. Now, Cl^- and K^+ ions migrate out of the middle region of the gel. In steady state a very thin zone of pure water is formed between the acidic and alkaline regions of the gel. In this neutral zone $[H^+] \approx [OH^-] \approx \sqrt{K_w} = 10^{-7}$, where K_w is the ionic product of water. For large backward voltages most of the voltage drop occurs in this high impedance zone of pure water. Based on the above assumptions, another formula was derived [12] for high backward voltages:

$$|i| = F \cdot (D_H + D_{OH}) \cdot \frac{2c_0 + \sqrt{K_w}|\Delta \varphi|}{l}. \qquad (2)$$

Now, the H^+ and OH^- ions carry most of the electric current and the contribution of K^+ and Cl^- ions is negli-

gible. In this case the gel can be divided into three parts. There is an alkaline region in the gel where the current is carried mainly by OH^- ions. The next is a thin neutral zone, where both OH^- and H^+ ions contribute to the current. Finally, there is the acidic region of the gel where H^+ ions carry most of the current. According to this simplified picture, recombination of H^+ and OH^- ions takes place at the two interfaces of the three different regions mentioned above. When experimental current–voltage characteristics measured on PVA based hydrogel cylinders (see Fig. 4) were compared to the theoretical ones a good agreement was found in the forward direction [12]. In the backward direction, however, the measured current and the slope of the current–voltage curves were orders of magnitude higher than expected. An investigation to find out the reason for this disagreement initiated the present work. Our experiments reported here show that fixed anions contaminating the gel are responsible for the observed high backward currents.

Experimental section

Chemical

Polyvinyl alcohol (nominal molecular weight 15 000, Fluka), glutardialdehyde (25% aqueous solution, grade I, Sigma), Cab–O–Sil fumed silica (grade H-5, Cabot Corp.), poly(diallyl dimethyl ammonium chloride) (20% solution, molecular weight 240 000, Polysciences), poly(styrene sulphonic acid, sodium salt) (completely sulphonated, molecular weight 500 000, Polysciences), p-dimethylamino benzaldehyde (Reanal), acetic acid (99.5% solution, Reanal) were used without further purification. All other chemicals were of reagent grade.

Gel recipe

0.2 g Cab–O–Sil was thoroughly mixed with 2 ml distilled water (or 2 ml of a polyelectrolyte solution when gels with fixed ionic groups were prepared) and with 1 ml glutardialdehyde solution (2% diluted from the stock). The mixture was degassed with an aspirator to prevent bubble formation in the later steps of the procedure. Then 4 g of polyvinyl alcohol solution (28.7% w/w) was added and the mixture was homogenized by stirring with a glass rod. Finally, 1 ml 5 M HCl was added under continuous stirring. The gelling of this mixture is rather fast, thus it was used immediately to prepare the gel cylinders. Cab–O–Sil was applied in the recipe because the mechanical strength and the stability of the gel was greatly improved this way.

Progr Colloid Polym Sci (1996) 102:101–109
© Steinkopff Verlag 1996

Gel cylinders

A few ml of freshly prepared liquid gel mixture was filled into a 5 ml plastic syringe and with the aid of the syringe a silicon rubber tubing (inner diameter 0.6 mm, outer diameter 2 mm, length 60–70 mm) was filled with the mixture. After 30 min the silicon tubing containing the gel cylinder was immersed into n-hexane, where the silicon tubing swelled considerably. To remove the gel cylinder from the tubing a syringe filled with n-hexane was attached to the tubing. Then, pressing the syringe forced the hexane to flow through the swollen tube and the gel cylinder slipped out of the tube simultaneously. The gel cylinders were placed into distilled water for 1 h and then into 1 M KOH for another hour. Gel cylinders prepared this way were stored in distilled water.

Inserting the gel cylinders into a hole

A small circular hole (diameter 0.34 mm) was cut into a 0.32-mm-thick Teflon membrane with the method described previously [12]. Before placing the gel cylinder into the hole the gel should be shrunk. To this end, the gel cylinder was placed into acetone for 2 min to substitute most of its water content by the volatile organic solvent. Then, the gel cylinder was placed into a dry Petri dish where the acetone evaporated and the gel shrank. After 1 h, the dry and somewhat hardened gel stick was cut into approximately 2-mm pieces. Such a piece was placed into the hole of a Teflon membrane. Then, a drop of distilled water was placed to both ends of the gel. Swelling of the gel closed the hole completely forming a "plug" (see Fig. 2a). Finally the Teflon membrane with the gel "plug" was placed into a 90 °C water bath for 1 hour. After this treat-

ment the Teflon discs with the gel were stored under distilled water.

Modified gels

Gels with fixed anionic groups were prepared by substituting the 2 ml distilled water with another 2 ml aqueous solution of 0.02 or 0.2 equivalent/l poly(styrene sulphonic acid, sodium salt) in the standard gel recipe (concentrations are given for the ionic groups). Gels with fixed cationic groups were prepared in a similar way but using 2 ml 0.2 eq./l or 20% (w/w) poly(dimethyl diallyl ammonium chloride) solutions. Dimethylamino groups were fixed in the PVA gels in the following way. Before the treatment the gel cylinder stood in glacial acetic acid for 10 min. Then the cylinder was placed again for 10 min into a 10% (w/w) solution of paradimethylamino benzaldehyde dissolved in glacial acetic acid. In the next step the gel was transferred into a 1:1 mixture of glacial acetic acid and 1 M HCl where it stood for another 10 min. Finally, the gel was washed and stored in distilled water.

Apparatus

A schematic picture showing the main units of the experimental setup can be seen in Fig. 1. The gel cylinder "G" is in the middle of the apparatus. (Details of the parts surrounding the gel are shown in Fig. 2.) The current-voltage characteristic was recorded with a four-electrode potentiostat controlled by a personal computer. The current electrodes ("I") were made of platinum wire. The voltage electrodes ("U") were Ag/AgCl electrodes in 1 M KCl solution ("K"). To suppress all liquid junction

Fig. 1 Experimental set-up used to measure polarization curves of the electrolyte diode. See text for further explanation

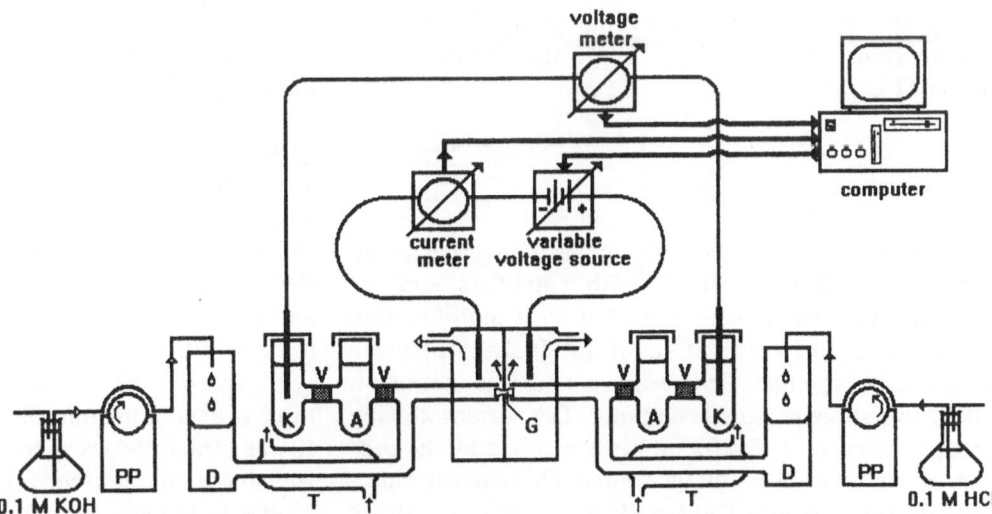

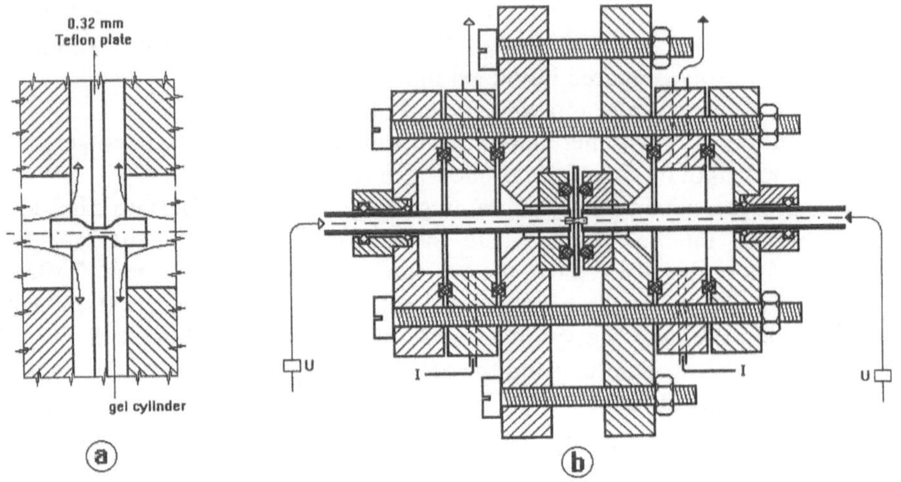

Fig. 2 Cross-sectional view of the central part of the apparatus. a) The gel cylinder in the Teflon membrane. b) Other parts of the apparatus. All parts were made of plexiglass except for the screws which were produced from hard PVC

Fig. 3 Theoretical polarization curve of an "ideal" pure gel (no fixed ions in the gel). Effective length: 1 mm. Electrolytes: 0.1 M KOH and 0.1 M HCl. Temperature: 25 °C. As formulae (1) and (2) predict only asymptotic behavior for large forward and backward voltages, the characteristic is drawn with a dashed line for small voltages

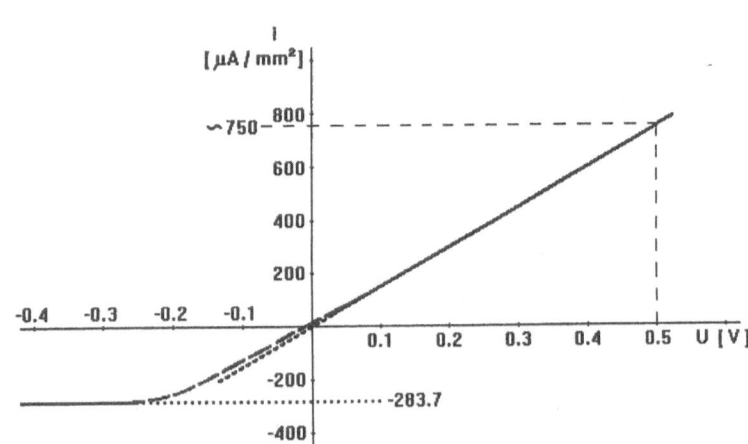

potentials, except in the gel, salt bridges filled with 10 M NH_4NO_3 ("A") were applied. ("V" denotes porous Vycor glass.) A steady flow (about 2 ml/min) of 0.1 M KOH and 0.1 M HCl was produced by peristaltic pumps ("PP"). The droppers ("D") were inserted to isolate any electric noise generated by the pumps. The flows were thermostatted to $25 \pm 1°C$ by the thermostating jackets "T".

Data collection

The measured current and voltage was converted to an electric signal between 0–5 V with amplifiers and attenuators. This signal was measured by a multifunction analog/digital I/O card (PC-LabCard PCL-711S) using the Labtech NotebookPro program and was stored on a 486-compatible personal computer. The current flowing through the current electrodes was controlled by the program across a variable voltage source. This current established the predetermined voltage drop between the voltage electrodes. The computer program monitored the voltage drop and deviations from the predetermined values were corrected via this feedback loop. The variable voltage source was an isolation amplifier and it was controlled by the analog output of the PC-LabCard.

Results and discussion

First, we studied a polyvinyl alcohol gel cylinder without any added polyelectrolyte. The current-voltage characteristic of such a "pure" gel cylinder is depicted in Fig. 4. It can be seen that the slope of the current-voltage curve in the reverse direction is much higher than expected for an "ideal" gel (see Fig. 3). The measured relative slope is $2.13 \cdot 10^{-3}$, while the theoretical value is $s_R = \sqrt{K_w}/2c_0 = 5 \cdot 10^{-6}$. That is, the measured value is nearly three orders of magnitude larger than the theoretical one. If was straightforward to assume that this phenomenon was due to the presence of some unidentified fixed ionic groups in the gel.

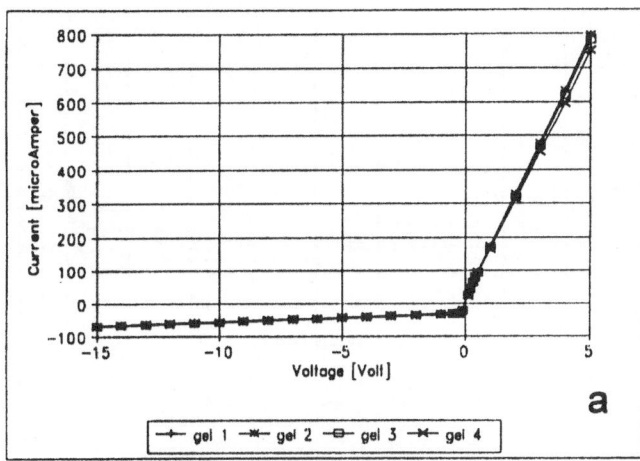
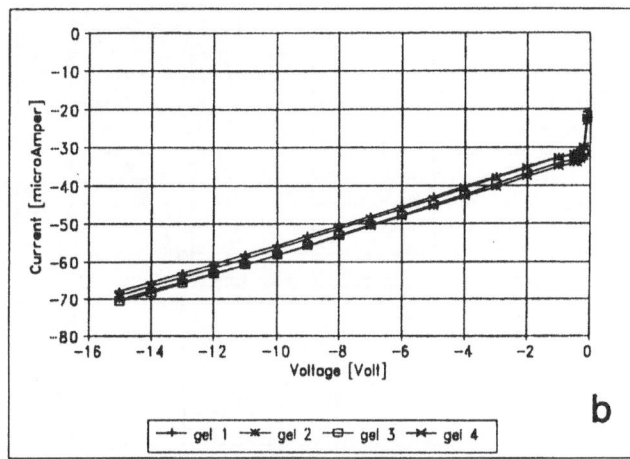

Fig. 4 Polarization curves of 4 "pure" gel cylinders. The electrolyte concentration c_0 was 0.1 M in all the experiments reported here. (a) Complete polarization curves. b) Backward characteristics (enlarged)

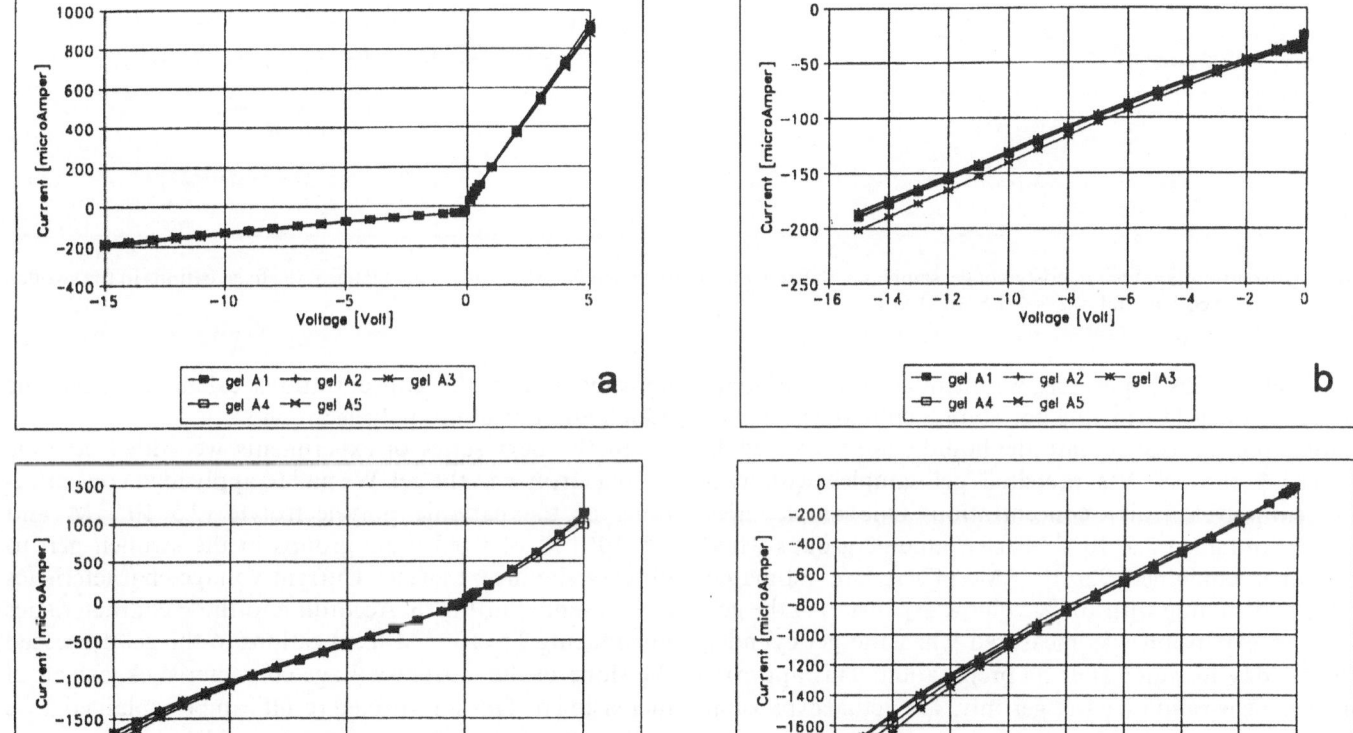

Fig. 5 Polarization curves of gel samples containing fixed ions of the anionic type. a) and b) c_{FA} (conc. of the fixed anions in the swollen gel) = $2.5 \cdot 10^{-3}$ eq/l. c) and d) $c_{FA} = 2.5 \cdot 10^{-2}$ eq/l

106
L. Hegedüs et al.
Reaction, diffusion, electric conduction and determination of fixed ions in a hydrogel

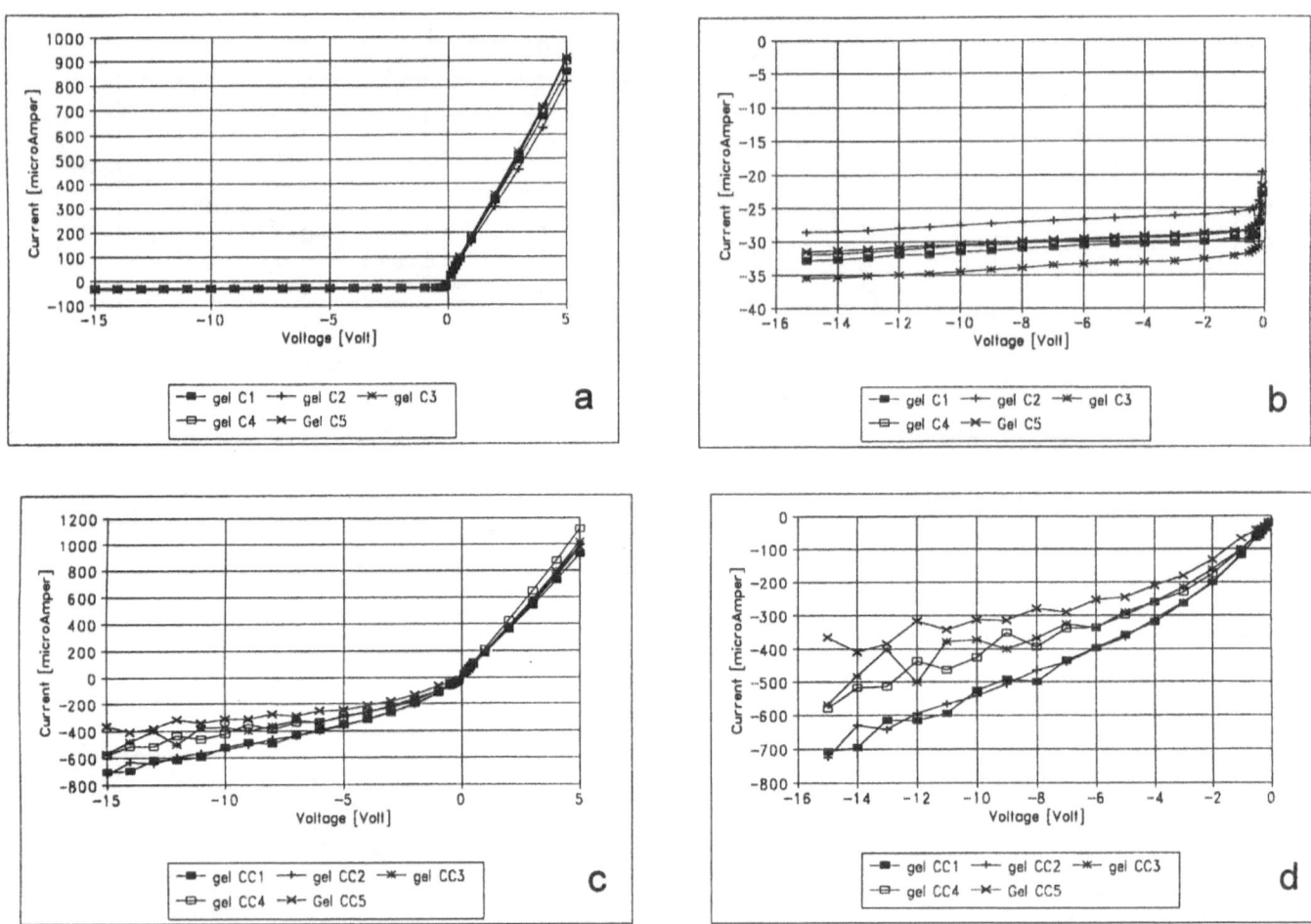

Fig. 6 Current-voltage characteristics of gel samples with fixed ions of the cationic type. a) and b) c_{FC} (conc. of the fixed cations in the swollen gel) = $2.5 \cdot 10^{-2}$ eq/l. c) and d) $c_{FC} = 1.5 \cdot 10^{-1}$ eq/l

To check this hypothesis, we introduced fixed anionic groups to the gel by mixing polystyrene sulphonic acid (an anionic polyelectrolyte) into the liquid components while preparing the gel. We prepared gel samples with two different polyelectrolyte concentrations. One sample series of gels contained $2.5 \cdot 10^{-3}$ M fixed anionic groups while the other contained $2.5 \cdot 10^{-2}$ M. (These are estimated values calculating with 100% volume expansion of the gel. Such an expansion was measured when the gel cylinder spent 1 day in water after its preparation. This approximate value is valid for a free gel only. The actual expansion of the gel samples used in our experiments was probably somewhat smaller because it was restricted by the Teflon membrane. Thus, the calculated concentrations give only a low estimate.) The current-voltage characteristics of such anionic gels are shown in Fig. 5. From the relative slopes we were able to calculate the concentration of the fixed ionic groups using the theory given in the Appendix. According to the calculations these concentrations are $2.7 \cdot 10^{-3}$ M and $2.6 \cdot 10^{-2}$ M, which are in rather good

agreement with the concentrations calculated from the added amount of the polyelectrolyte.

In the next series of experiments we added cationic polyelectrolyte to the gel. We had to apply high concentrations of the cationic polyelectrolyte ($2.5 \cdot 10^{-2}$ M and $1.5 \cdot 10^{-1}$ M of fixed ionic groups in the swollen gel) to observe significant effects. Current-voltage characteristics are presented in Fig. 6. According to these characteristics introducing $2.5 \cdot 10^{-2}$ M fixed cations to the gel decreased the slope of the current-voltage characteristics instead of increasing it. This unexpected result can be explained if we assume that the unidentified "intrinsic" fixed ionic groups of the gel are anionic in nature. These intrinsic anionic groups can neutralize the effect of the added cationic polyelectrolyte. Actually a salt is formed in the gel where both the anion and cation is a polyelectrolyte. As there are less mobile counterions in this case the current in the gel is decreased. Experiments applying $1.5 \cdot 10^{-1}$ M fixed cationic groups in the gel gave high backward currents. Steady currents were found only for low voltages (between

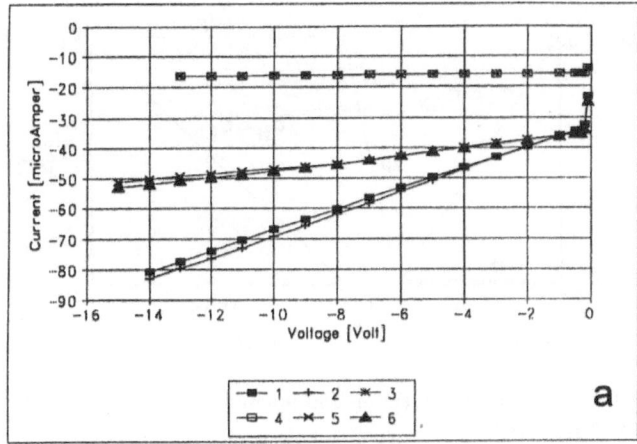 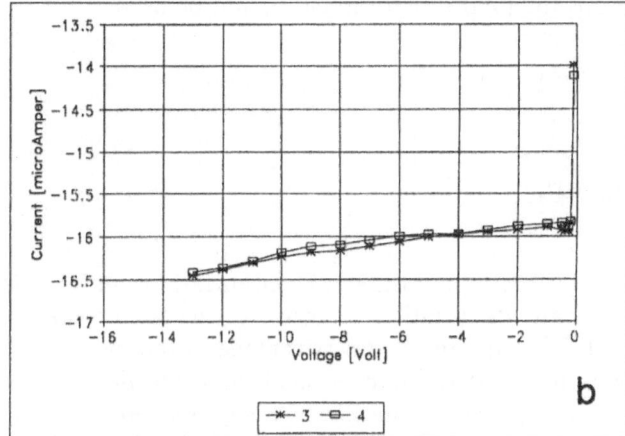

Fig. 7 Polarization curves of two gel samples with fixed dimethylamino groups (backward direction only). a) curve 1, 2: current-voltage characteristic before fixing the groups, curve 3, 4: characteristic one hour after the fixation, curve 5, 6: polarization curve two days after the fixation; b) curve 3, 4 enlarged to show the slope for the curve

0 and 5 V) and above 10 V currents oscillations were observed in some cases. In addition, a drift of the measured currents was detected, e.g., after 1 day the current at 15 V decreased from 700 μA to 400 μA. A possible explanation for this is that the applied polycation was not stable in the conditions of our experiments. Another possibility is that the observed effects are due to some slow migration of the polyelectrolyte out of the high electric field zone of the gel.

All the results discussed previously suggest that our "pure" gel is contaminated with some fixed anionic groups. Thus a backward characteristic approximating the ideal one can be observed only if these contaminating groups are neutralized somehow. To this end we fixed weak cationic groups to the gel in excess. We used para-dimethylamino benzaldehyde for this purpose. The aldehyde group reacts with the PVA fixing the dimethyl-amino groups to the polymer network. The current-voltage characteristic of such a gel is shown in Fig. 7. Just as expected the relative slope of this gel is nearly two orders of magnitude smaller than that of the untreated "pure" gel. The fixation of the weak cationic groups with the above method is not long lasting, however. After a few days of standing the slope of the backward characteristic increases again as it is shown in Fig. 7.

We tried to identify the source of the contaminating anionic groups in the gel. First, we suspected that the glutardialdehyde was contaminated with some glutardial-dehyde acid. However, preparing gels with very pure samples of glutardialdehyde produced no measurable difference in the backward currents. Thus the unknown fixed anionic groups should be associated with the PVA component of the gel. The contaminant most probably is

a weak acid. This assumption is supported by the following considerations. Assuming that the contaminant is a strong acid its concentration calculated from the relative slope should be about $7 \cdot 10^{-4}$ M. In this case adding $2.5 \cdot 10^{-2}$ M cationic groups to the gel – as we did in our experiments – should cause high backward currents. According to our results, however, this amount did not increase but rather even decreased the backward current. This result suggests the presence of a weak acid in a higher concentration. Experiments are in progress to identify this unknown component of the gel.

Conclusion

The most important conclusion of the present work is that when gels are used in an electrolyte diode experiment, then the concentration of the ionic groups fixed in the gel can be determined simply from the relative slope of the backward characteristic. This concentration is given by the following formulae:

$$c_{FA} = \frac{2c_0 \cdot (D_H + D_{OH})}{D_H} \cdot s_R \qquad (3)$$

$$c_{FC} = \frac{2c_0 \cdot (D_H + D_{OH})}{D_{OH}} \cdot s_R \qquad (4)$$

if the gel with the fixed groups forms a strong cation or anion exchanger, respectively. (See expressions (A12) and (A14) in the Appendix.) In the case of weak ion exchangers it is only a fraction of the fixed groups which is ionized. Thus to calculate the total concentration of the fixed groups the following formulae should be used:

$$c_A = \frac{c_{FA}}{\sqrt{K_A}} \qquad (5)$$

$$c_B = \frac{c_{FC}}{\sqrt{K_B}}, \qquad (6)$$

where c_A and c_B are the total concentrations and K_A and K_B are the dissociation constants of the fixed weak acid and base, respectively. An advantage of the method presented here is that relatively small concentrations of fixed ionic groups can be measured easily. Moreover, for the determination of the fixed ionic group concentration in a gel no calibration is needed as the relative slope method uses the intercept as an inner standard. No information is needed about the size of the gel sample (length and cross section) either. The only data necessary for the calculation – beside the relative slope – is c_0, that is, the concentration of the acid and the base used in the experiments. This concentration can be varied to find a situation where both the slope and the intercept of the backward characteristic can be determined accurately.

Acknowledgement This work was supported by OTKA grants F 007572 and T 017041.

Appendix

Backward current of the electrolyte diode
in the presence of fixed ions

As an approximation let us assume one-dimensional concentration $c_i = c_i(x)$ and potential $\varphi = \varphi(x)$ profiles in the gel cylinder. (See Fig. 8.)

The derivation starts with the Nernst–Planck equations:

$$\frac{J_i}{D_i} = -\frac{dc_i}{dx} - z_i c_i \frac{d\varphi}{dx} \qquad i = H^+, OH^-, K^+, Cl^- \qquad (A1)$$

and assumes electroneutrality in the gel:

$$\sum z_i c_i = 0. \qquad (A2)$$

J_i is the current density, D_i is the diffusion coefficient, c_i is the concentration, z_i is the charge of the i-th ion and φ is the dimensionless potential. The electric current density for the alkaline, the acidic and the middle zone of the gel can be calculated separately knowing Δx_{OH}, Δx_H and δ respectively.

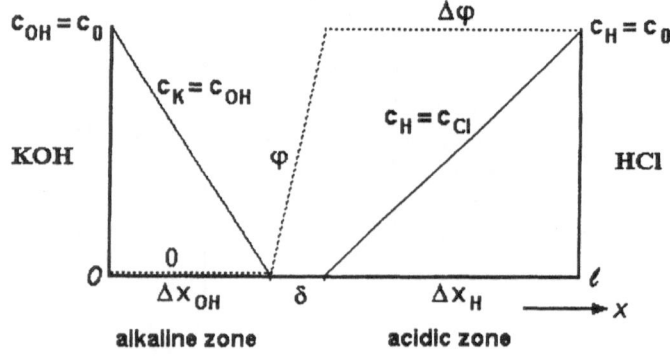

Fig. 8 Schematic concentration and potential profiles for large backward voltages

In the alkaline region

$$\frac{J_K}{D_K} + \frac{J_{OH}}{D_{OH}} = -2\frac{dc_K}{dx} = const = \frac{2c_0}{\Delta x_{OH}}, \qquad (A3)$$

because of the electroneutrality condition. Δx_{OH} is the length of the alkaline zone. For large backward voltages J_K is very small compared to J_{OH}. Thus, the absolute value of the current density in this region is

$$|i| = F \cdot |J_{OH}| = F \cdot D_{OH} \cdot \frac{2c_0}{\Delta x_{OH}}. \qquad (A4)$$

In a similar way for the acidic region

$$\frac{J_{Cl}}{D_{Cl}} + \frac{J_H}{D_H} = -2\frac{dc_{Cl}}{dx} = const = -\frac{2c_0}{\Delta x_H}, \qquad (A5)$$

and

$$|i| = F \cdot |J_H| = F \cdot D_H \cdot \frac{2c_0}{\Delta x_H}. \qquad (A6)$$

Finally, in the middle zone of length δ fixed ions and its counterions can be found in an approximately constant concentration. If the fixed ions are anions then the H^+ concentration in this zone can be expressed as

$$c_H \approx c_{FA}, \qquad (A7)$$

assuming that the concentration of other ions is negligible in this region. c_{FA} is the concentration of the fixed anions. The absolute value of the current density in this region is

$$|i| = F \cdot |J_H| = F \cdot D_H \cdot c_{FA} \cdot \frac{|\Delta \varphi|}{\delta}, \qquad (A8)$$

taking into account that $|J_H| \gg |J_{OH}|$. It is also assumed that the total voltage drop is limited to the middle zone.

The sum of the length of the three zones is equal to the length of the gel:

$$\Delta x_{OH} + \delta + \delta x_H = l \,. \tag{A9}$$

Combining the above expression and the ones found for the three separate zones, the following equation can be derived:

$$\frac{|i|}{F} = \frac{2c_0(D_H + D_{OH}) + D_H \cdot c_{FA} \cdot |\Delta\varphi|}{l} \,, \tag{A10}$$

as $|i|$ should be the same in all the three zones in a steady state. Thus the expression of the slope/intercept or relative slope s_R is

$$s_R = \frac{D_H \cdot c_{FA}}{2c_0(D_H + D_{OH})} \tag{A11}$$

and c_{FA} can be calculated from the measured s_R value as

$$c_{FA} = 2\frac{D_H + D_{OH}}{D_H} \cdot s_R \cdot c_0 \,. \tag{A12}$$

In the case of fixed cations an analogous derivation yields the following formulae:

$$\frac{|i|}{F} = \frac{2c_0 \cdot (D_H + D_{OH}) + D_{OH} \cdot c_{FC} \cdot |\Delta\varphi|}{l} \tag{A13}$$

and

$$c_{FC} = \frac{2c_0 \cdot (D_H + D_{OH})}{D_{OH}} \cdot s_R \,, \tag{A14}$$

where c_{FC} is the concentration of the fixed cations.

References

1. Nosztyiczius Z, Horsthemke W, McCormick WD, Swinney HL, Tam WY (1987) Nature 329:619–620
2. Kshirsagar G, Nosztyiczius Z, McCormick WD, Swinney HL (1991) Physica D 49:5–12
3. Lázár A, Nosztyiczius Z, Farkas H, Försterling HD (1995) CHAOS 5:443–447
4. Castets V, Dulos E, Boissonade J, De Kepper P (1991) Phys Rev Letters 64:2953–2965
5. Ouyang Q, Swinney HL (1991) Nature 352:610–612
6. Nosztyiczius Z, Ouyang Q, McCormick WD, Swinney HL (1992) J Phys Chem 96:6302–6307
7. Lee KJ, McCormick WD, Swinney HL (1994) Nature 369:215–218
8. Lee KJ, McCormick WD, Nosztyiczius Z, Swinney HL (1992) J Chem Phys 96:4048–4049
9. Watzl M, Münster AF (1995) Chem Phys Lett 242:273–278
10. Lengyel I, Epstein IR (1991) Science 251:650–652
11. Kárpáti-Smidróczki É, Büki A, Zrínyi M (1995) Colloid and Polymer Sci 273:857–865
12. Hegedüs L, Nosztyiczius Z, Papp Á, Schubert A, Wittmann M (1995) ACH-Models in Chemistry 132:207–224
13. Nosztyiczius Z, Schubert A (1973) Periodica Polytechnica 17:165–177
14. Schubert A, Nosztyiczius Z (1977) Periodica Polytechnica 21:279–283

Progr Colloid Polym Sci (1996) 102:110–117
© Steinkopff Verlag 1996

Experimental and theoretical investigation of static and dynamic chemical pattern formations in gels

A. Büki
É. Kárpáti-Smidróczki
K. Meiszel
M. Zrínyi

Dr. A. Büki (✉) · É. Kárpáti-Smidróczki
K. Meiszel · M. Zrínyi
Technical University of Budapest
Department of Physical Chemistry
1521 Budapest, Hungary

Abstract Static and dynamic Liesegang structures have been studied experimentally and by means of computer simulation. Reaction diffusion differential equations of the system were solved by method of finite differences. Equations were based on Ostwald's supersaturation theory and a modified sol coagulation model. It was possible to reproduce the normal as well as the revert type of Liesegang phenomenon. In the case of dynamic structures effect of the initial concentrations and structure of the polymer matrix have been studied.

Key words Liesegang phenomenon – reaction-diffusion system – gel – simulation

Introduction

In the last three decades pattern-forming chemical, physical and biological processes came into the foreground of scientific interest. The two main causes of this are the exotic nature of these processes and – in the last decade – the dramatic evolution of computer techniques. The latter is important because it has made possible to solve numerically the reaction-diffusion differential equations by which the above-mentioned processes can be described and which generally cannot be handled by analytic mathematical methods.

The most commonly known and studied example of chemical pattern formation is the so-called Belousov-Zhabotinsky reaction which is a chemical oxidation of malonic acid by bromate with a very complex mechanism. Formation of a regular structure is due to the coupling of chemical and diffusion processes, not only in this case, but generally whatever chemical reaction the pattern formation is based on.

If we want to understand the basic facts of pattern-forming processes investigation of a rather simple chemical reaction is more useful, because of the small number of components and experimental parameters.

If a precipitate forming chemical reaction between two electrolytes takes place in a gel matrix in a diffusion limited way, one can often find a pattern of precipitate consisting of bands or stripes depending on the geometry of the experimental set-up. This is the so-called Liesegang phenomenon which was discovered more than a century ago, but has never been exactly described mathematically. Although this pattern formation is based on a very simple reaction, the reaction–diffusion differential equations of the system cannot be solved by means of analytic mathematical methods.

The most simple and general way to perform a Liesegang experiment is to use a gel cylinder or thin layer as a diffusion medium. One of the two reacting electrolytes is contained by the gel and the solution of the other one is poured onto the top of the cylinder or a drop of its concentrated solution is placed onto the gel film.

The main role of the gel is to prevent convection of solutions and sedimentation of the precipitate. The only way for the transport of the electrolytes is diffusion, i.e., the whole process is diffusion limited. Some representative examples of the static Liesegang structures can be seen on Fig. 1.

In some special cases the precipitate can redissolve in the excess of the outer electrolyte. If this process is fast

Progr Colloid Polym Sci (1996) 102:110–117
© Steinkopff Verlag 1996

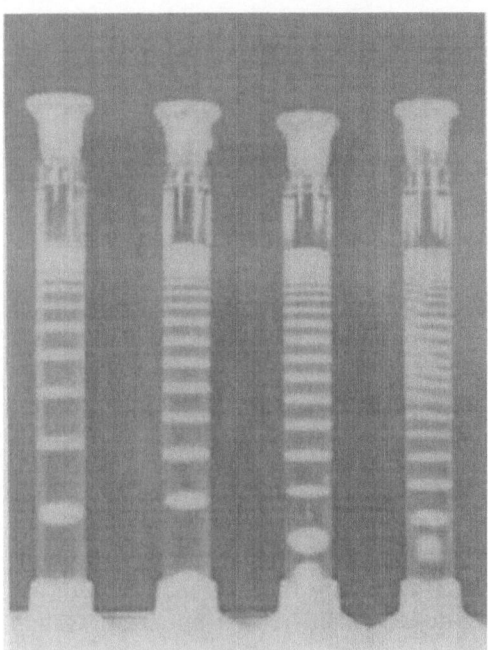

Fig. 1 Some representative examples of the static Liesegang structures

enough a moving precipitate band will be formed instead of a static structure. The movement however is apparent because on one side of the band precipitate is formed continuously, whereas on the other side it continuously redissolves. We have named this phenomenon "heterogenous travelling wave" because it is very similar to chemical waves occurring in oscillating chemical reaction systems [1].

In our work, we have studied static and dynamic Liesegang structures experimentally and by means of computer simulation.

Static structures

Theoretical background

Static Liesegang patterns are regular both in time and space, which is expressed by the following two scaling laws [2].

Distances between adjacent precipitate zones are members of a geometrical series which has a quotient higher than one. This means that the length of clear spaces between the zones increases in the direction of diffusion. In the literature the quotient is called spacing coefficient (P), which is expressed by Jablczinsky's spacing law:

$$P = \frac{X_{n+1}}{X_n} , \tag{1}$$

where X_n and X_{n+1} are the distances between the nth and $(n + 1)$th zones, measured from the gel surface. If the zones have a significant thickness, the distances are measured from the middle of the zone.

The ratio of the distance of a zone measured from the gel surface and the square root of time elapsed until the zone appears is constant:

$$X_1 : X_2 : \cdots : X_n = \sqrt{t_1} : \sqrt{t_2} : \cdots : \sqrt{t_n} . \tag{2}$$

This is often regarded as the time law of the Liesegang phenomenon.

Patterns that satisfy both of the above-mentioned requirements are referred to as "regular" Liesegang structures. However, in some cases, one can find unusual patterns such as spirals, spheres or highly ramified tree-like structures.

Experimental

We have performed our experiments in chemically cross-linked poly(vinyl-alcohol) (PVA) gels. Gels were prepared by cross-linking of primary PVA chains with glutaric aldehyde (GDA) in aqueous solution. Commercial PVA (Merck 821038) and solution of 25 wt% GDA (Merck) were used. The initial polymer concentration as well as the cross-linking density could be altered. This latter means the ratio of monomer unit (VA) to the cross-linking agent (GDA). This ratio [VA]/[GDA], called degree of cross-linking, was varied between 50 and 400.

In order to study the precipitation in swollen networks, one of the reactants was mixed with the polymer solution containing the cross-linking agent. Then slow gelation process was induced by decreasing the pH of the system by nitric acid (Carlo Erba). After that the solution was poured into glass tubes or onto glass plates. The size of the tubes as well as that of the plates were also varied. At pH = 2 the gelling process took place within 5–7 h. After completion of the network formation, the gels were brought into contact with the reactant, allowing it to diffuse into the gel.

Discussion of our experimental results can be found in our earlier papers [1, 3].

Theories of the Liesegang pattern formation

The Liesegang phenomenon was discovered more than a century ago, however, there is not a comprehensive theory which could prove all the effects and anomalies described in the literature.

There are two main theories for description of the one-dimensional regular Liesegang phenomenon [2]. One of them is the so-called supersaturation theory by Ostwald which was the first model of the Liesegang pattern formation. This regards formation of the precipitate as a kinetically hindered autocatalytic process. Formation of the crystals has a double barrier. The product of electrolyte concentrations has to reach a critical value, the so-called solubility product which is a thermodynamically parameter of the reaction system. The other barrier is caused by a kinetical effect. In almost every system formation of the first crystals needs higher concentration of the electrolytes than the growth of existing precipitate. This kinetical barrier vanishes after formation of first crystals, which means that the crystallisation is an autocatalytic process.

According to this model, formation of the precipitate starts only at relatively high supersaturation but after that continues with a very high speed. Because of this high reaction speed the slow diffusion cannot feed the reaction, which means that sharp precipitate bands and depleted zones, that is, a pattern will form in the system.

The other often applied theory is the so-called sol coagulation model. According to this model the precipitate forms at first as a stable sol, which can even diffuse in the system by a smaller speed than the electrolytes. When the concentration of the outer electrolyte reaches a certain value (the so-called critical coagulation concentration) stability of the sol vanishes, particles will aggregate and because of their large size they will not be able to move in the network any more. At the place of the coagulation a visible precipitate zone will appear. Around the zone concentration of the sol particles decrease which causes slow diffusion of the sol towards the precipitated zone. This leads to formation of a depleted zone around the precipitate band where formation of the sol particles can continue, but the sol concentration is too low for coagulation. This means that an alternating pattern of stable and coagulated precipitate will form.

From an experimental point of view the main difference between the two theories is the distribution of the precipitate. According to the supersaturation theory areas between precipitate zones are really empty, that is, they do not contain precipitate at all. Against this the sol coagulation model leads to a continuous precipitate distribution. The zones between bands are only virtually empty because they contain a dilute stable sol.

Computer simulation of possible mechanisms

The build-up of a Liesegang pattern includes two main processes: diffusion of the components and the reaction between them. Formation of a pattern is caused by the coupling of these two processes.

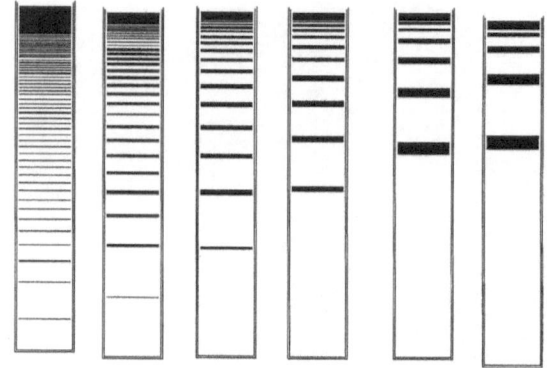

Fig. 2 Some simulated one-dimensional Liesegang patterns

Both processes can be described by exact mathematical equations. In most simple cases for description of diffusion Fick's law can be applied. For kinetics of nucleation and crystal growth there are many theories in the literature, too. The reaction-diffusion differential equation system, however, cannot be solved by the usual analytic methods.

This insufficiency of exact mathematics makes computer simulation a very useful method in this field [4–7]. In our work, we have studied the two above mentioned theories by means of numerical models. Goal of our work was to establish mechanisms by means of which not only the regular Liesegang phenomenon can be described, but the anomalous structures mentioned in the literature and found by us can be proved.

Results with the supersaturation model

We have applied a one-dimensional model for this mechanism in which we have solved Fick's differential equation by the well known method of finite differences [8, 9]. The program has determined amount of the precipitate at every point of the diffusion column in every step according to the above described qualitative model. Precipitation was regarded as an irreversible process and the precipitate was AB type, i.e., it contained only one ion of every electrolytes. Some simulated one-dimensional patterns can be seen in Fig. 2. More details about the simulation procedure can be found in [3].

The simulated patterns have fulfilled both the time and spacing laws mentioned above. We have performed many simulations with different parameters and tried to determine their influence for the spacing coefficient. Varied parameters were the following:

- concentrations of inner and outer electrolytes
- diffusion coefficients
- length of diffusion column
- precipitation and nucleation products.

Progr Colloid Polym Sci (1996) 102:110–117
© Steinkopff Verlag 1996

Fig. 3 Demonstration of the
Matalon–Packter equation

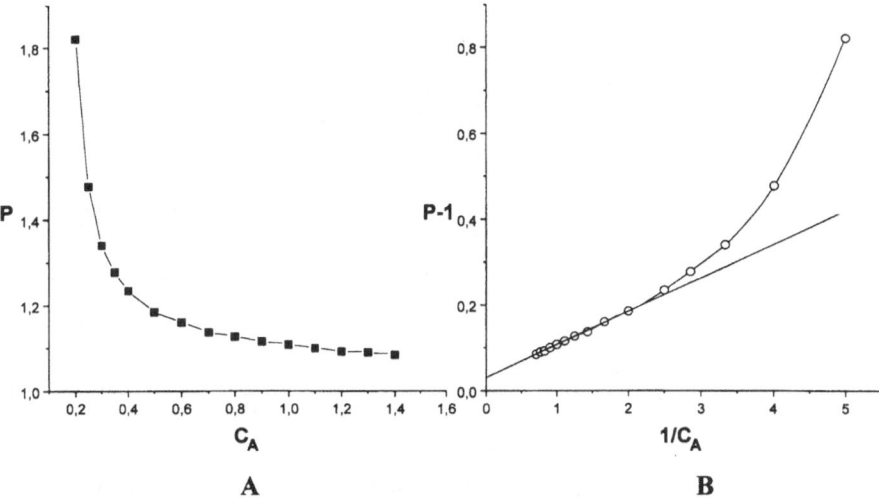

A

B

Generally, the found dependencies cannot be described by simple analytic functions. The only exception was the diffusion coefficients. The spacing coefficient depends linearly on the diffusion velocity of the inner electrolyte. In case of the outer electrolyte, we have found a reciprocal dependence.

From the dimensional analysis of the problem it follows that spacing coefficient is exactly determined by the following four reduced parameters:

$$D_R = \frac{D_B}{D_A} \qquad S = \frac{K_s}{L},$$

$$S_0 = \frac{L}{C_A C_B} \qquad A = \frac{C_A - C_B}{C_A + C_B},$$

where D_R is the ratio of the diffusion coefficients, S is the degree of supersaturation, S_0 is the initial supersaturation at the junction point and A characterises the asymmetry of initial concentrations. However, except for D_R, it was not possible to exactly determine their influence on the spacing coefficient [4].

In the Liesegang experiments the most frequently studied parameters are the concentrations of the inner and outer electrolytes. Apart from these and the composition of the gel matrix there are no other important parameters which can be easily controlled by the chemist. Matalon and Packter [10] have found the following simple relation between the concentration of the outer electrolyte and the spacing coefficient:

$$P - 1 = A + \frac{B}{C_A}, \tag{3}$$

where C_A is the concentration of the outer electrolyte. We have not found this equation to be generally valid. The transformation prescribed by the above equation is shown in Fig. 3a, but the plot is more close to exponential dependence than to linear. However, the region of this curve that corresponds to high outer electrolyte concentrations can be fitted by a linear function (see Fig. 3b). So on the basis of computer simulation we may conclude that the Matalon–Packter relation is valid only for cases when the concentration of the outer electrolyte is high enough.

Anomalies

Validity of the scaling law can be checked for a given structure by plotting the logarithmic distances of the zones versus the ordinal number of them. If this function is linear the structure is a regular Liesegang pattern and the slope of the linear function is the spacing coefficient.

In some cases the upper end of the above-mentioned function falls out of the linearity. This deviation can show positive as well as negative tendency.

In some linear patterns which were formed in chemically cross-linked poly(vinyl-alcohol) gels from $Mg(OH)_2$ and were described in an our previous paper [3] we have found a negative deviation. This means that the increasing of the distances between the zones is something slower at the end of the structure than at the beginning. In other words, the structure can be characterised by two spacing coefficients because the $Ln(X_n)$ vs. n function consists of two linear parts with different slopes.

It was possible to reproduce this anomaly by the supersaturation model. The only thing which had to be done was to find the proper set of input parameters. If the concentration and diffusion coefficient of the outer electrolyte was chosen to be higher by about one order of magnitude than the same parameters of the inner electrolyte the above-mentioned irregularity appeared in the structure.

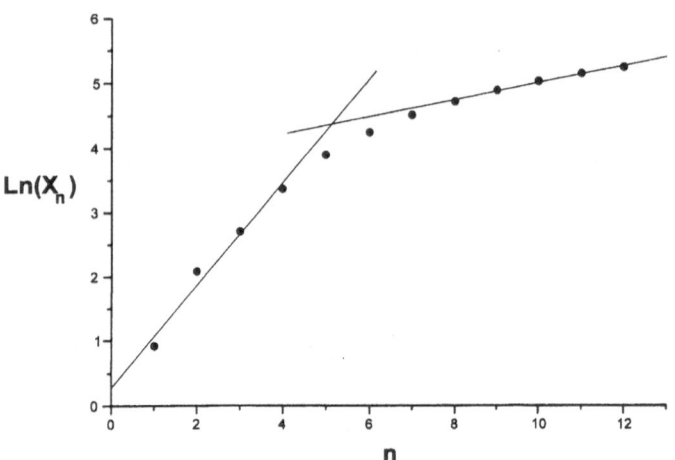

Fig. 4 $Ln(X_n)$ vs. n function of an anomalous simulated pattern

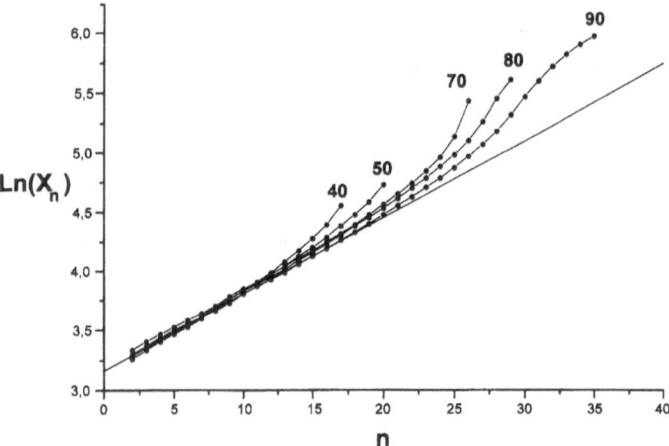

Fig. 6 $Ln(X_n)$ vs. n function for some simulated patterns

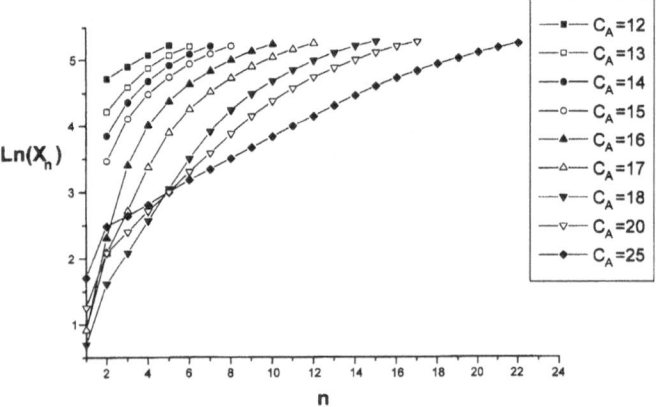

Fig. 5 The position of the breaking point of the one-dimensional anomalous structures which contain negative deviation can be controlled by the concentration of outer electrolyte

This ratio of the mentioned parameters is in a good agreement with the reality because the hydroxide ions can diffuse about ten times faster than other ones and, according to our experiments, one can find pattern formation only when the concentration of the outer component is much higher than that of the inner electrolyte. Figure 4 shows the $Ln(X_n)$ vs. n function of an anomalous simulated pattern.

We have found that the position of the breaking point can be altered by the concentration of the outer electrolyte as is shown in Fig. 5.

The positive deviation from the linearity was described by Holba [11]. He has observed this anomaly in a radially developed $Ag_2Cr_2O_7$ system. Jablczinsky's relation was found to be valid not only for linearly but for radially arranged Liesegang patterns so this anomaly must not be a geometric effect.

We were able to reproduce the mentioned deviation by taking into account the depletion of the outer electrolyte during the development of the structure. In other ways the applied mechanism was the same as the previously described one-dimensional supersaturation model. The volume of the outer electrolyte measured in arbitrary units was added to the initial parameters. The algorithm was continuously controlling the amount of outer electrolyte flowing into the diffusion column, and subtracting it from the initial amount has determined a new outer electrolyte concentration in every step according to the following equation:

$$C_A(0, t_j) = \frac{1}{V}\left(V \cdot C_{A0} - \sum_i (C_A(X_i, t_{j-1}) + P(X_i, t_{j-1}))\right).$$

(4)

Here, V is the volume of the outer solution measured in arbitrary units, C_{A0} is the initial concentration of the electrolyte, $C_A(X_i, t_j)$ is the concentration at the point X_i and time t_j and $P(X_i, t_j)$ is the amount of the precipitate.

Figure 6 shows the $Ln(X_n)$ vs. n function for some simulated patterns. The numbers next to the curves are the volume of the outer electrolyte measured in arbitrary units. As can be seen, the smaller this volume is the closer the anomalous deviation occurs to the beginning of the structure.

Sol coagulation model and revert structures

In the literature there are many papers which concern anomalous Liesegang structures [2]. The most prevalent anomaly mentioned by other workers is the so-called revert Liesegang structure. This pattern consists of

Progr Colloid Polym Sci (1996) 102:110–117
© Steinkopff Verlag 1996

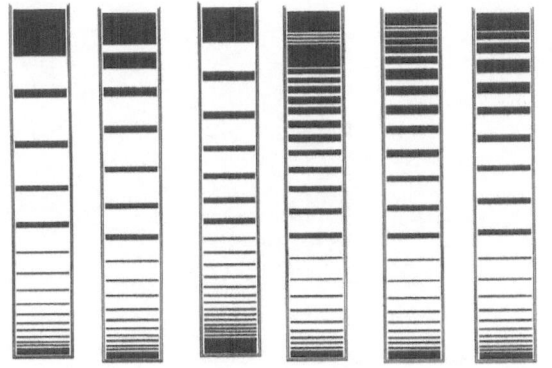

Fig. 7 Some one-dimensional patterns simulated by means of the modified sol coagulation model

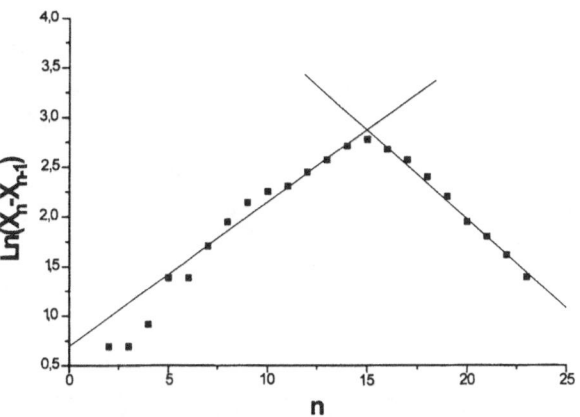

Fig. 8 Determination of spacing coefficients for the regular and revert sections of an anomalous simulated pattern

precipitate bands just like the regular one, however, distances between them decrease in the direction of diffusion.

According to our findings, regular structures can be reproduced exactly by means of the supersaturation model, whereas this mechanism cannot lead to evolution of a revert structure. This finding allows us to conclude that the Liesegang phenomena found in different reaction systems cannot be described by only one comprehensive mechanism. That is why we have tried out in the second part of our work the other main mechanism, the sol coagulation model.

The main purpose of this work was to design experiments which can be realised in practice and by means of which it can be decided whether a certain system functions according to one or the other mechanism.

We have found that the sol coagulation model can lead to formation of a pattern only if the coagulation is autocatalytic. This was first advised by Shinohara [12] who has developed an analytic description for the Liesegang phenomenon based on the sol coagulation theory. In other ways the numerical model developed for this mechanism was the same as the one applied for the supersaturation model, except that here the precipitate was able to diffuse while it was in a sol state.

In most of the cases the sol coagulation model resulted in revert Liesegang patterns. However, with proper initial parameters it was possible to get regular structures by means of this model. Some simulated patterns are shown in Fig. 7. It can be seen that in many cases double structure forms which consists of a normal and a revert section. The first is always the regular one which has a spacing coefficient higher than one. The end of the patterns is a revert section with a spacing coefficient smaller than one (Fig. 8). Position of the junction point between them can be controlled by the initial conditions.

Dynamic structures, experimental results

In case of certain chemical reactions a weakly soluble salt is produced, which can redissolve in the surplus of one of the reactants due to complex formation. An example of such reaction is formation of $Cr(OH)_3$:

$$Cr(NO_3)_3 + NaOH = \underline{Cr(OH)_3} + 3NaNO_3$$

$$\underline{Cr(OH)_3} + NaOH = Na[Cr(OH)_4]$$

The white $Cr(OH)_3$ precipitate can, to some extent, redissolve in the surplus of NaOH, providing soluble material of green colour, $Na[Cr(OH)_4]$.

In case of this precipitate, instead of expected static Liesegang patterns, dynamic systems have developed. At the beginning of experiments a zone of reaction occurs at the interface when reactants come into contact by diffusion. After a certain time, one observes not only advancing, but also receding boundary surface of precipitate. The width of this precipitate layer (between the advancing and receding interfaces) gradually increases up to a certain value and then it seems to remain constant. The precipitate band grows on one side and redissolves on the other, thus for the observer it seems to be moving. The propagating band, which we call a heterogeneous travelling wave, can be characterised by the following quantities:

– position of the propagating band as a function of time (l),
– width of the precipitate layer and its dependence on time (d),
– stability of the propagating band which is defined as the longest path which a travelling precipitate can achieve before it disappears.

We found that a moving band keeps on slowing down until it disappears. The longest distance that a $Cr(OH)_3$

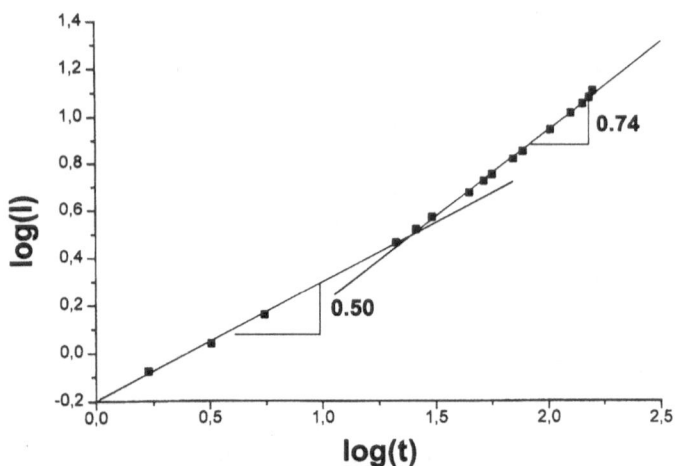

Fig. 9 The log(l)–log(t) dependence of a moving band can be approximated by two straight lines

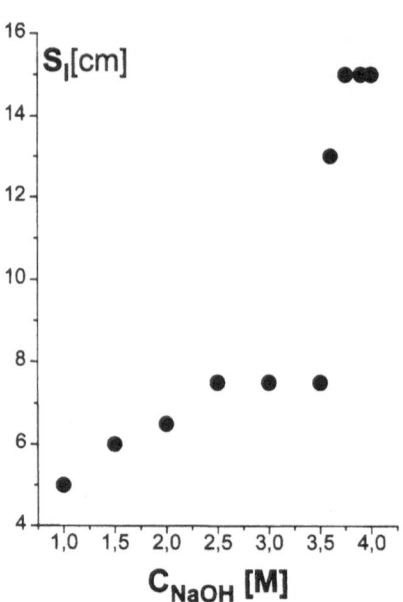

Fig. 10 Dependence of stability of a moving band on the concentration of outer electrolyte

precipitate layer passed was 130 cm. It is important to mention that in case of $Cr(OH)_3$ precipitate during the propagation the thickness of the bands was found to be constant within the experimental error. On the basis of experimental data, one may conclude that, in most cases, the movement of the band does not follow Fick's law of diffusion:

$$l = k t^{1/2} , \qquad (5)$$

where l denotes the displacement at time t and k represents a proportionality factor. We found that the displacement is very seldom proportional to $t^{1/2}$.

In many cases the kinetics of propagation can be divided into two parts. At the beginning, the movement of $Cr(OH)_3$ obeys the Fickian law, that is the displacement scales with $t^{1/2}$. For longer times, the power law dependence remains valid, but the exponent α has been found to be greater than 1/2.

Figure 9 illustrates this finding. The log(l)–log(t) dependence can be approximated by two straight lines. The first one has a slope of 0.50 which corresponds to Fickian diffusion, whereas the second straight line can be represented by a significantly greater slope. There is a cross-over between the Fickian and non-Fickian behaviour about log(t) $\cong$ 1.3. Thus the kinetics of moving precipitate band can be described generally as follows:

$$l = \text{const} \cdot t^{\alpha} , \qquad (6)$$

where $\alpha = 0.5$ at small displacements and $\alpha \sim 0.7$ at long displacements.

It was established that polymer concentration of the gel does not significantly influence the velocity of movement. This means that the presence of polyvinyl alcohol chains has no effect on precipitation (including nucleation)

or on dissolution. It is worth to mention that in case of radial diffusion α was found to be smaller than 1/2.

The lifetime of propagating band strongly influenced by the PVA content. With increasing polymer concentration the stability significantly decreases. The cross-linking density does not make its influence felt on the stability.

With increasing concentration of inner electrolyte, the stability significantly increases, then above a certain concentration it decreases again. The dependence of stability on the concentration of outer electrolyte is shown in Fig. 10. It may be seen that in the concentration region 1.0–3.5 M NaOH a slight increase is experienced. However at a concentration 3.5 M $< C <$ 3.75 M a sharp increase in the stability can be observed. In this range the stability becomes twice as before. Not only the stability, but also the velocity of the propagating band changes significantly in this concentration range.

The temperature plays an important role in the stability. At higher temperature the moving precipitates are less stable. At 50 °C the moving layers exist only for 7 h, whereas at lower temperature, for example at 23 °C, the bands travel more than 10 times longer. Above 50 °C no travelling bands have been observed.

We can only speculate at present about the mechanism that could account for the above described phenomena. One of the most important tasks is to understand what kind of coupling exists between precipitation and dissolution which can maintain a constant thickness of the travelling wave.

Conclusions

Gels of chemically cross-linked poly(vinyl-alcohol) chains are excellent medium for studying the pattern forming precipitation in reaction-diffusion systems. We have developed a one-dimensional model based on Ostwald's supersaturation theory in order to simulate Liesegang pattern formation.

It was found that the Matalon–Packter equation agrees with our results only at high values of outer electrolyte concentration. Several deviations from the spacing law have been interpreted by taking into account depletion of the outer electrolyte during the reaction.

We were also able to show that on the basis of sol coagulation model revert Liesegang structures can be expected. Beside static precipitate patterns a dynamic phenomenon has been observed. A virtual propagation of $Cr(OH)_3$ precipitate band with constant thickness may be the consequence that the advancing interface is probably coupled with redissolution at the receding interface. Stability of the propagating band also supports this idea since the ceasing of both advancing and receding interface occurs at the same time.

Acknowledgement Our work was supported by the József Varga Foundation, and by the Hungarian Academy of Sciences OTKA No. T015754 and F 014023.

References

1. Zrínyi M, Gálfi L, Smidróczki É, Rácz Z, Horkay F (1991) J Phys Chem 95:1618
2. Stern KH (1954) Chem Rev 54:79
3. Kárpáti-Smidróczki É, Büki A, Zrínyi M (1995) J Colloid and Polymer Sci 273/9:857–865
4. Büki A, Kárpáti-Smidróczki É, Zrínyi M: J Chem Phys, in press
5. Henisch HK (1986) J Crystal Growth 76:279
6. Henisch HK, Garcia-Ruiz JM (1986) J Crystal Growth 75:195
7. Henisch HK, Garcia-Ruiz JM (1986) J Crystal Growth 75:203
8. Crank J (1985) The Mathematics of Diffusion (Clarendon Press, Oxford)
9. Chopard B, Droz M (1991) Journal of Statistical Physics 64:859
10. Matalon R, Packter A (1955) J Colloid Sci 10:46
11. Holba V (1989) Colloid and Polym Sci 267:456
12. Shouji Shinohara (1970) Journal of the Physical Society of Japan 29:1073

Progr Colloid Polym Sci (1996) 102:118–122
© Steinkopff Verlag 1996

Hydrogel medicinal systems of prolonged action

Yu. Samchenko
Z. Ulberg
N. Pertsov

Dr. Yu. Samchenko (✉) · Z. Ulberg
N. Pertsov
Institute of Biocolloidal Chemistry
National Academy of Sciences of Ukraine
85 Frunze str.
254080 Kyiv, Ukraine
tel: (044)435-30-09
(fax)

Abstract The copolymerization of acrylamide with acrylic acid has allowed to synthesize therapeutic contact lenses, first of all, antiglaucoma ones. The reached considerable prolongation of the release of miotic reagent pilocarpine from such lenses (and other polymer carriers) is explained by the formation of the ionic bonds between polyionic hydrogel and medicine substance being nitrogen base. Slowly running hydrolysis of the formed ionic complex is accompanied by gradual release of the drug.

Sharp change in an amount of the water, retained by the hydrogels, at small pH change allows to control the process of transport and release of the medicines introduced in the polymer carrier.

Key words Hydrogel – copoly-merization – pilocarpin – ophthalmology – prolonged release

Introduction

Hydrogels are polymers characterized by hydrophilicity and insolubility in water [1]. They are quite perspective from the view-point of their use in medicine, especially in ophthalmology [2–5]. At present, the soft contact lenses, production of which reaches tens of millions pieces per year, belong to the most wide-spread products using hydrogels for medical purposes [6].

Besides sight correction, often the necessity of dosed introduction of various medicines (first of all, while treating glaucoma) appears in ophthalmologic practice [7].

Traditionally used in ophthalmology, the instillation method for medicine administration has a number of substantial shortcomings, the main ones of which can be the necessity for multiple repeated procedure during the day and the absence of stable maintenance of the required drug concentration in the eye tissue. Therefore, it is clear that the problem of the immobilization of medicines in soft contact lenses (or other polymer carriers) with subsequent prolonged release is extraordinarily important. Few examples of the immobilization of medicines, antiglaucoma pilocarpine preparation in particular, by the soft contact lenses with 38% water content on the base of 2-hydroxyethyl methacrylate are described in the literature [8]. It is noted that inspite of low water content, which must promote the diffusion retardation, about 70% of medicine and practically all pilocarpine amount is washed off within 1 h and 3 h respectively.

Still faster, practically complete wash-off of pilocarpine is observed in the case of homopolyacrylamide contact lenses with 85% equilibrium water content, which provide satisfactory feeding of cornea with oxygen.

As far as is known, the two factors listed below most substantially affect the rate of diffusion of the drugs in hydrogels: i) the content of the solvent in macromolecular network and ii) the energy of the interaction of diffusing compound with the macromolecular network. Because of the fact that the sharp decrease of the equilibrium water content in hydrogels is undesirable as it causes the decline of oxygen permeability, the only way to retard the release of a medicine from the polymeric carrier is to intensify its interaction with macromolecular network.

The copolymerization of acrylamide with acrylic acid has allowed to solve this problem and synthesize on their base therapeutic contact lenses, first of all, antiglaucoma ones. The reached considerable prolongation of the release of miotic reagent pilocarpine from such lenses is explained by the formation of the ionic bonds between polyionic hydrogel and medicine substance being nitrogen base. Slowly running hydrolysis of the formed ionic complex is accompanied by gradual release of the drug.

Experimental section

Materials

Acrylamide (AM) and acrylic acid (AA) were purified according to conventional methods [9, 10].

N,N′-Methylenebisacrylamide (cross-linker, MBA) were purchased from SERVA and used without further purification, as well as potassium persulphate and sodium metabisulfite (redox system).

Synthesis

Homopolyacrylamide gels (PAAG) and copolymer gels on the base of AM and AA were prepared by free-radical polymerization in water at 298 K according to the method reported previously by the authors [11].

Measurements

The ability of hydrogels to swell in water and in standard buffer solutions with different pH values was studied by the gravimetric method. Water content of the samples was determined by the following equation:

$$M(g/g) = \frac{\text{weight of water (in g) in polymer}}{\text{weight of dry polymer (in g)}}. \quad (1)$$

The water state in the swelled samples was studied by the method of the differential scanning calorimetry using the DSM-2M microcalorimeter [11]. Bound water content (W_{nf}) was determined by the following equation:

$$W_{nf} = W - W_f, \quad (2)$$

where W and W_f are respectively equilibrium water content and freezing water content.

Pilocarpine release from the swollen copolymer hydrogels was studied using spectrophotometer SPECORD M40 according to the method reported by Graham et al.

[14]. Maximum UV absorption of aqueous pilocarpine hydrochloride occurs at 215 nm.

Results and discussion

Specific interactions between pilocarpine and a polymer matrix are confirmed by the considerable concentrating effect influenced the pilocarpine solutions by the copolymer hydrogels based on acrylamide and acrylic acid (Fig. 1). We see that if in case of the homopolyacrylamide gel the distribution coefficient does not exceed 1.5, in case of the copolymer hydrogels it increases up to 7.5. Thus, in case of polyacrylamide gel, the greater part of medicine is incorporated into a polymer due to its swelling in the medicine solution. As to the copolymer hydrogels mentioned process will be supplemented by the selective sorption on the active carboxylic groups of the polymer macromolecule.

As is seen from Fig. 2 for all investigated hydrogels, the rate of release of the pilocarpine decreases in time. For the first minutes the rate of release is very high, especially in the case of the homopolyacrilamide gel – by the 5th minute it is about 7% per minute. With the increase of the content of the carboxylic groups in the hydrogels, bonding molecules of pilocarpine and gradually hydrolyzing with its release, the rate of release is substantially retarded.

Principal differences in the mechanism of the absorption and retention of pilocarpine by hydrogels containing carboxylic groups and by non-ionogeneous polyacrylamide gels also stipulate the difference in the kinetics of its release (Fig. 3). If in 1 h about 80% of sorbed pilocarpine was already released from homopolyacrylamide gel, about 20% was released by the mentioned moment from hydrogel containing 15% of the acrylic acid links and from the hydrogel containing 45% of carboxylic groups, only 6% was released and even after 20 h still about 75% against the initial pilocarpine amount was retained.

The found coefficients of the pilocarpine diffusion also demonstrate the principle differences between the homopolyacrylamide gel and copolymer hydrogels containing active carboxylic groups. In the case of the latter the pilocarpine diffusion coefficients are lower approximately by two orders ($10^{-6} - 10^{-7}$ and $10^{-8} - 10^{-9}$ cm²/s respectively) and the prolonged preparation release is the consequence of this.

In the semi-logarithmic coordinates shown in Fig. 4, the time dependences of diffusion coefficients of the pilocarpine for the copolymer gels based on AM and AA have a distinctive shoulder, up to which the ordinary wash-off of a medicine, having no links with polymeric matrix by the specific interactions, makes the prevalent contribution in the release of pilocarpine, and after – the wash-off with the previous hydrolysis of ionic bonds

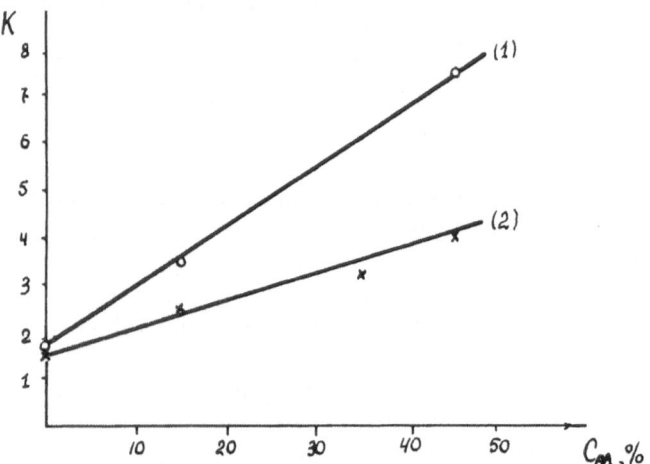

Fig. 1 Dependence of the distribution coefficient of the pilocarpine in the system hydrogel–water for the hydrogels with different monomer content $(1 - C_{MBA} - 0.84; 2 - C_{MBA} = 0.42)$

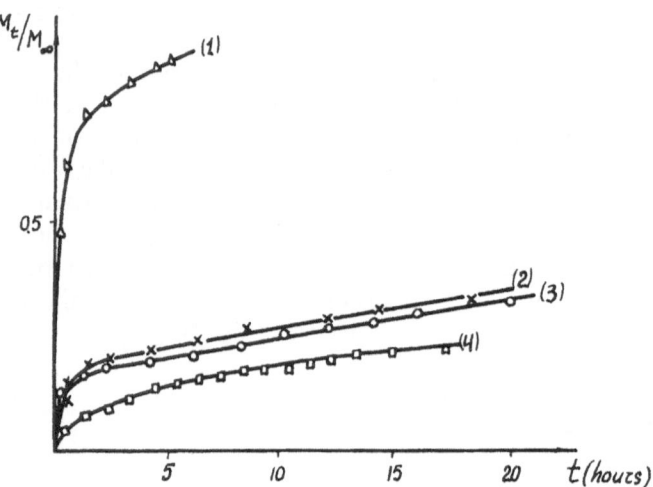

Fig. 3 Kinetics of the pilocarpine release for hydrogels with different composition: $1 - $ PAAG; $C_{MBA} = 0.42\%$; $2 - C_{AA} = 15\%$; $C_{MBA} = 0.42\%$; $3 - C_{AA} = 35\%$; $C_{MBA} = 0.21\%$; $4 - C_{AA} = 45\%$; $C_{MBA} = 0.84\%$

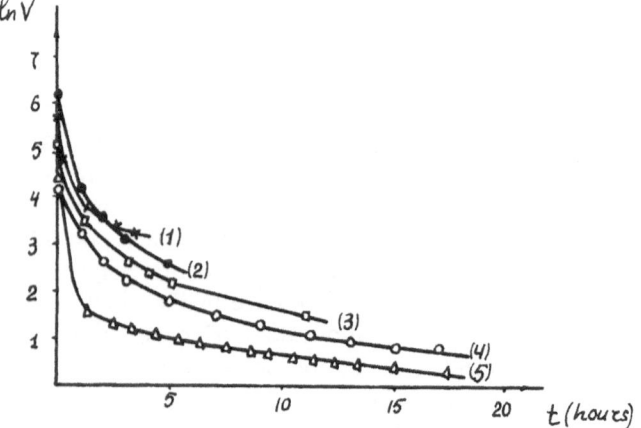

Fig. 2 Time dependence of the rate for the pilocarpine release for various hydrogel systems $(1 - $ PAAG, $C_{MBA} = 0.42\%$; $2 - $ PAAG, $C_{MBA} = 0.84\%$; $3 - C_{AA} = 15\%$; $C_{MBA} = 0.42\%$; $4 - C_{AA} = 35\%$, $C_{MBA} = 0.42\%$; $5 - C_{AA} = 45\%$, $C_{MBA} = 0.84\%)$

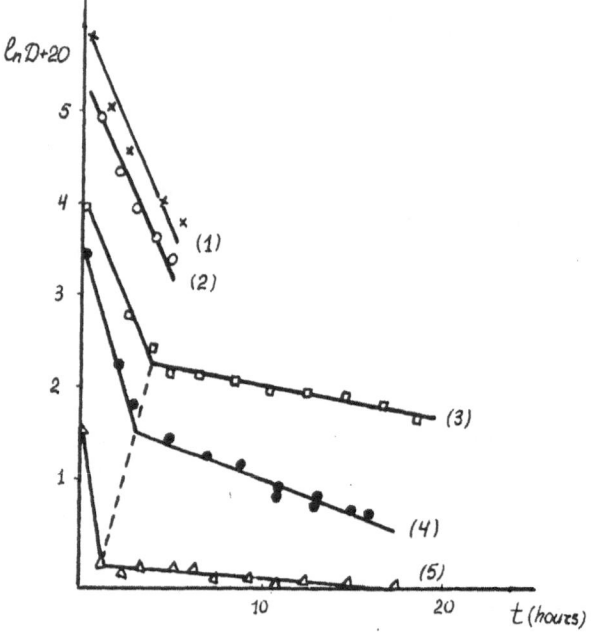

Fig. 4 Time dependences of the coefficients of the pilocarpine diffusion for different hydrogel systems: $1 - $ PAAG; $C_{MBA} = 0.42\%$; $2 - $ PAAG; $C_{MBA} = 0.84\%$; $3 - C_{AA} = 15\%$; $C_{MBA} = 0.84\%$; $4 - C_{AA} = 35\%$; $C_{MBA} = 0.84\%$; $5 - C_{AA} = 45\%$; $C_{MBA} = 0.84\%$

between the molecules of pilocarpine and carboxylic groups, going with substantially lower rate. As the conditional line, drawn through the points of the shoulder, shows – the higher content of ionic groups in the polymeric matrix, the earlier starts the realization of the second mechanism of the two mechanisms of release described here. So far as the polyacrylamide gel is concerned, the wash off from it goes according to the first mechanism, and this is explained by the absence of chemical interaction between the medicine preparation and the polymeric matrix.

The rate of diffusion of medicine substances from gels containing 60–90% of water cannot but depend upon the state of water in them. The influence of the composition of copolymer hydrogels based on acrylamide and acrylic acid

with different concentrations of the cross-linking agent upon the state of water in them is shown in Fig. 5. As it is seen from the figure, the growth of the bound water content reaching 45% was observed during the partial substitution of acrylamide links in the polymer matrix by the links of acrylic acid. The freezing water content in this case decreased. The growth of the non-crystallizing water in

Progr Colloid Polym Sci (1996) 102:118–122
© Steinkopff Verlag 1996

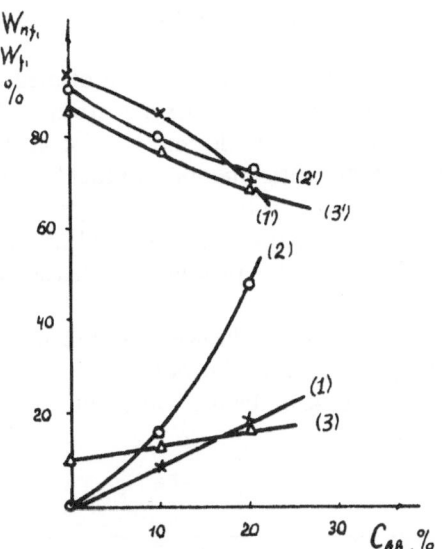

Fig. 5 Dependence of the bound water content (W_{nf},%; 1, 2, 3) and the freezing water content (W_f,%; 1', 2', 3') on the AA content in the reaction mixture for the various concentrations of the cross-linking agent: 1, 1' – 0.037%; 2, 2' – 0.094%; 3, 3' – 0.225%

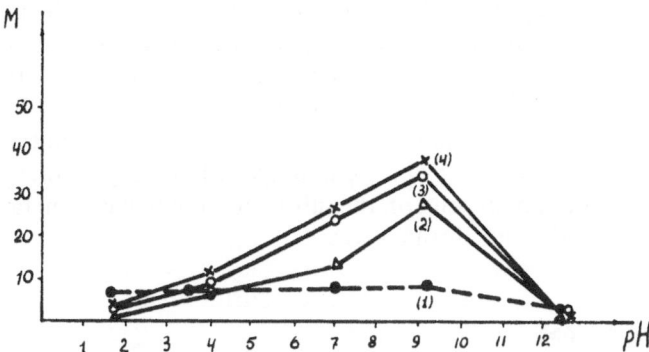

Fig. 7 Equilibrium water content as function of pH value for various hydrogel systems (1 – PAAG; 2 – $C_{AA} = 40\%$; 3 – $C_{AA} = 80\%$; 4 – $C_{AA} = 100\%$; $C_{MBA} = 0.63\%$)

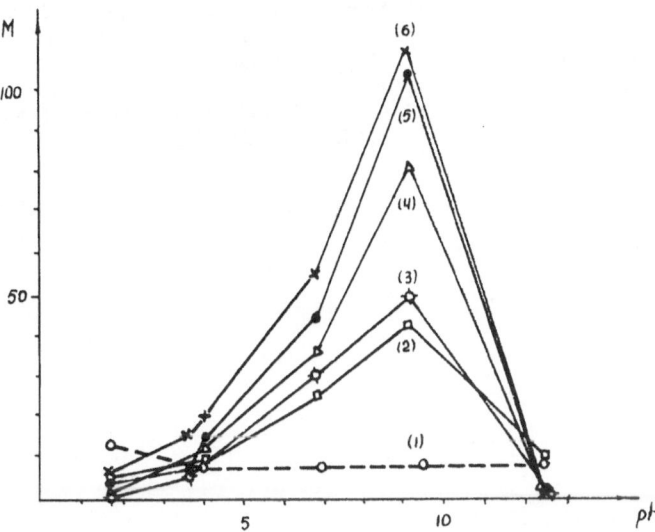

Fig. 6 Equilibrium water content as function of pH value for various hydrogel systems (1 – PAAG; 2 – $C_{AA} = 20\%$; 3 – $C_{AA} = 40\%$; 4 – $C_{AA} = 60\%$; 5 – $C_{AA} = 80\%$; 6 – $C_{AA} = 100\%$; $C_{MBA} = 0.066\%$)

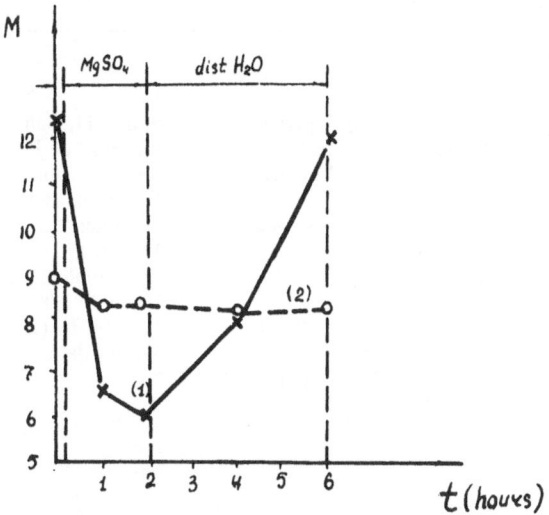

Fig. 8 Swelling behaviour of homopolyacrylamide gel and copolymer hydrogel, containing carboxylic groups (1 – PAAG; 2 – $C_{AA} = 100\%$; $C_{MBA} = 0.21\%$)

copolymer hydrogels as compared with the polyacrylamide gel indicates that the water in them is subjected to stronger influence of the polymer matrix.

The possibility to use hydrogel carriers in stomatology is closely connected with their ability to sharply change the swelling degree with slight pH change. We investigated the dependence of the acrylamide–acrylic acid copolymer hydrogels swelling degree (as well as the amount of a liquid retained by 1 g of dry polymer, M) on pH using six standard buffer solutions in the pH range from 1.68 to 12.45

(Fig. 6). As is seen, the homopolyacrylamide gels swelling practically does not depend on the medium pH, but in case of copolymer hydrogels the increase in the swelling degree reaches the maximum in the pH alkaline region.

The range of the change in the swelling degree is considerably narrowed with the growth of the cross-linking agent (MBA) content that is explained by a densely packed structure of the hydrogels in closely cross-linked state and increased with the growth of the carboxylic groups content (Fig. 7).

Sharp decrease in the swelling degree of the copolymer hydrogels at pH 12.45 (saturated Ca(OH)$_2$ solution) can be explained by the formation besides covalent cross-linkings of ionic ones in a medium of bivalent metals that results in the volume transition in collapsed phase.

With placing a polyelectrolyte hydrogel in 2% MgSO$_4$ solution for 2 h the quantity of the liquid retained by it

122

Yu. Samchenko et al.
Chemical modification of acrylamide gels

decreases by more than two times as compared with the swelling in distilled water (Fig. 8). Repeated placing of a collapsed sample in distilled water for 4 h causes growth of its swelling degree (and thus the permeability for the incorporated substances in a gel) to return to the initial value. The swelling of a homopolyacrylamide gel practically does not depend on the substitution of bivalent metal salt solution for distilled water.

Conclusions

It has been shown that copolymerization of acrylamide with acrylic acid allows to influence upon the physico–chemical properties of produced gels over a wide range.

The considerable retardation of the diffusion of medicine preparations, retained by the carboxylic groups of copolymeric gels based on acrylamide and acrylic acid, enables to produce from them antiglaucoma contact lenses combining the corrective effect together with the hypotensive action upon the eye of a patient and as well as the other polymeric carriers with the prolonged release of medicine substances.

Sharp change in the amount of the water retained by the hydrogels, at small pH change, allows to control the process of transport and release of the medicines introduced in the polymer carrier.

References

1. Wichterle O (1987) In: Encyclopedia of Polymer Science and Engineering. John Wiley & Sons, New York, Vol 7, pp 783–807
2. Chiellini E, Saettone M (1988) J Bioact Compat Polym 3(1):86–93
3. Sastre R, Mateo J (1988) Rev Plast Mod 55(379):77–83
4. Madruga E, San Roman J (1987) Rev Plast Mod 54 (377):675–681
5. Saettone M, Torracca M, Pagano A (1992) Int J Pharm 86:159–166
6. Singer H, Bellantoni E (1987) In: Encyclopedia of Polymer Science and Engineering, John Wiley & Sons, New York, Vol 4, pp 164–173
7. Salminen L (1987) FIDIA Res Ser 11:161–170
8. Zatloubal Z, Doleral P (1986) Cesk Farm 35:318–321
9. Maurer H (1968) Disk-Electrophorese. Walter de Gruyter. Berlin
10. Nemec J, Bauer W (1985) In: Encyclopedia of Polymer Science and Engineering. John Wiley & Sons, New York, Vol 1, pp 211–234
11. Samchenko Yu, Atamanenko I, Baranova A, Ulberg Z, Altshuler M (1991) Dopovidi Akademii nauk Ukrainskoi RSR 6:127–129
12. Graham N, Zulfiqar M (1988) J Controlled Release 5:243–252

Progr Colloid Polym Sci (1996) 102:123–125
© Steinkopff Verlag 1996

M. Dragusin
D. Martin
M. Radoiu
R. Moraru
C. Oproiu
S. Marghitu
T. Dumitrica

Hydrogels used for medicine and agriculture

Dr. M. Dragusin (✉) · D. Martin
M. Radoiu · R. Moraru
C. Oproiu · S. Marghitu · T. Dumitrica
Institute of Atomic Physics
IFTAR-Electron Accelerator Laboratory
P.O. Box MG-6
Magurele
76900 Bucharest, Romania

Abstract Three types of hydrogels obtained by electron beam irradiation and gamma ray irradiation, homopolymers of acrylamide (pAAm type), co-polymers of acrylamide-natrium acrylate (pAAmNa type) and homopolymers of natrium acrylate (pNaAc type) are presented. The effects of radiation absorbed dose and swelling medium nature upon the swelling degree and mechanical strength of these types of hydrogels are also discussed. Their proper physical characteristics can be well controlled by chemical composition of the solution to be treated by electron beam irradiation and by a suitable adjustment of the radiation absorbed dose. The best values obtained for the mechanical strength and swelling degree in distilled water for these hydrogels are, respectively: 25 kPa and 10 g/g for pAAmNa type, 6 kPa and 180 g/g for pNaAc type, 50 kPa and 8 g/g for pAAm type. These values are different from the ones obtained by gamma ray irradiation. Our experimental results proved that both the swelling degree and the mechanical strength depend on radiation absorbed dose level and swelling medium nature.

Key words Hydrogels – radiation – electron

Introduction

The hydrogels, such as pAAm type (homopolymer of acrylamide), pAAmNa type (co-polymer of acrylamide-natrium acrylate) and pNaAc type (homopolymer of natrium acrylate), were developed for the following applications: in agriculture to maintain soil humidity, to obtain a loose soil, to be used as fertilizer for dry soil (pAAm and pAAmNa); in rubber vulcanization (pAAmNa); in medicine as absorption material for dressing (pNaAc). Our radiation research in the hydrogels field started with IETI-10,000 Co[60] sources at the Institute of Nuclear Physics and Engineering (INPE) and then transferred at ALIN-10 and ALID-7 linacs, built in the Institute of Physics and Technology for Radiation Devices (IPTRD).

Radiation research in the field of hydrogels

Preparation of pAAm, pAAmNa and pNaAc hydrogel types [1, 2] is based on polymerisation, by gamma and electron beam irradiation, of the aqueous solutions of acrylamide, acrylamide-natrium acrylate and natrium acrylate, respectively. The radical reaction mechanism of such solutions polymerisation depends on the total monomer concentration (TMC) as well as on the water presence in the system. Furthermore, the radicals originated from irradiated water have a predominant role on the radicals which come directly from the monomers' irradiation. For the irradiated solutions in which the water concentration decreases under 40%, the conversion coefficient decreases, although the absorbed dose is increased.

The polymerisation of the above-mentioned aqueous solutions may be influenced by the following factors: chemical composition, radiation absorbed dose (D) level and radiation absorbed dose rate (d) level. The typical chemical compositions of the aqueous solutions to be irradiated are:

— 40% TMC for pAAm hydrogel type;
— 40% TMC, 90% acrylamide and 10% sodium acrylate, for pAAmNa hydrogel type;
— 30% TMC for pNaAc hydrogel type.

In the present work, concerning these hydrogels, we had in view the influence of the solution chemical composition, swelling medium nature (SMN) and absorbed dose upon two important characteristics: swelling degree (SD) and mechanical strength (MS). As swelling mediums we have used distilled water, physiological serum (0.09% aqueous solution of NaCl) and NaCl aqueous solution of normality 4 (4 N NaCl). The mechanical strength was tested by crossing the hydrogel granules through a sieve with 800 μm holes using a HS-02 device type (IGH-Hungary).

Table 1 presents the influence of D and SMN upon swelling degree SD of pAAm hydrogel type, obtained by electron beam irradiation. As indicated in Table 1, due to the pAAm neutral character, SMN has a small effect upon SD at the same D, and D has an important effect upon SD only when SMN is distilled water. In the case of 4N NaCl aqueous solution as swelling medium, D has a small effect upon the swelling degree of the pAAm hydrogel type.

Table 2 gives the influence of D and SMN upon mechanical strength of pAAm hydrogel type. The results presented in Table 2 indicate that the mechanical strength of pAAm hydrogel type increases for D from 4 kGy to 8 kGy, by a factor of 4.4 in the physiological serum case and by a factor of 5.6 in distilled water case. For 4N NaCl aqueous solution as swelling medium, D has a small effect upon MS.

The optimum value of D giving good results for both SD and MS is 8 kGy for the pAAm hydrogel type.

Table 3 shows the effect of D and SMN upon SD of pAAmNa hydrogel type, obtained by electron beam irradiation. As indicated in Table 3, the absorbed dose level has an important effect upon SD only in the distilled water case: SD at D of 16 kGy is about 3.5 times smaller than SD at D of 4 kGy.

Table 4 presents the effect of D and SMN upon mechanical strength of pAAmNa hydrogel type. In this case, the experimental results show a small dependence of MS versus D. From technological point of view, the optimum value for D giving good results for both SD and MS is 8 kGy for the pAAmNa type case.

Table 1 The effect of radiation absorbed dose (D) and swelling medium nature (SMN) upon the swelling degree (SD) of the pAAm type hydrogel (pAAm type = homopolymer of acrylamide)

D (kGy)	4	8	16
SD (g/g) SMN: distilled water	12.7	9.1	6.0
SD (g/g) SMN: physiological serum	12.7	9.2	8.5
SD (g/g) SMN: 4N NaCl aqueous solution	8.3	8.1	7.8

Table 2 The effect of radiation absorbed dose (D) and swelling medium nature (SMN) upon mechanical strength (MS) of the pAAm type hydrogel (pAAm type = homopolymer of acrylamide)

D (kGy)	4	8	16
MS (kPa) SMN: distilled water	11	62	81
MS (kPa) SMN: physiological serum	29	133	116
MS (kPa) SMN: 4N NaCl aqueous solution	78	87	101

Table 3 The effect of radiation absorbed dose (D) and swelling medium nature (SMN) upon the swelling degree (SD) of the pAAmNa type hydrogel (pAAmNa type = co-polymer of acrylamide-natrium acrylate)

D (kGy)	4	8	16
SD (g/g) SMN: distilled water	368	204	106
SD (g/g) SMN: physiological serum	27.5	24.4	23.5
SD (g/g) SMN: 4N NaCl aqueous solution	18.3	14.6	9.8

Table 4 The effect of radiation absorbed dose (D) and swelling medium nature (SMN) upon mechanical strength (MS) of the pAAmNa type hydrogel (pAAmNa type = co-polymer of acrylamide-natrium acrylate)

D (kGy)	4	8	16
MS (kPa) SMN: distilled water	6	8	9
MS (kPa) SMN: physiological serum	26	20	40
MS (kPa) SMN: 4N NaCl aqueous solution	46	75	43

Table 5 gives the effect of D and SMN upon SD of pNaAc hydrogel type obtained by electron beam irradiation. As indicated in Table 5, SD of pNaAc hydrogel presents a small dependence versus D, for all SMN types.

Table 5 The effect of radiation absorbed dose (D) and swelling medium nature (SMN) upon the swelling degree (SD) of the pNaAc type hydrogel (pNaAc type = homopolymer of natrium acrylamide)

D (kGy)	4	8	16
SD (g/g) SMN: distilled water	157	187	172
SD (g/g) SMN: physiological serum	140	135	150
SD (g/g) SMN: 4N NaCl aqueous solution	72	60	52

Table 6 The effect of radiation absorbed dose (D) and swelling medium nature (SMN) upon mechanical strength (MS) of the pNaAc type hydrogel (pNaAc type = homopolymer of natrium acrylamide)

D (kGy)	4	8	16
MS (kPa) SMN: distilled water	42	21	15
MS (kPa) SMN: physiological serum	6	4	3
MS (kPa) SMN: 4N NaCl aqueous solution	4	3	3

Also, SM of this hydrogel type presents a small dependence versus D. This conclusion results from experimental data presented in Table 6. Because pNaAc hydrogel is used in medicine as absorption material for dressing, its mechanical strength is not a critical one. So, the optimum value for D is 4 kGy.

The best value obtained by gamma irradiation for SD and MS are as follows:

– SD = 8 g/g and MS = 50 kPa for pAAm type;
– SD = 105 g/g and MS = 25 kPa for pAAmNa type;
– SD = 180 g/g and MS = 6 kPa for pNaAc type.

Conclusions

Our radiation research in the field of hydrogels proved that it is possible to obtain a very wide range of characteristics and therefore a large area of applications by varying irradiated solution chemical composition and absorbed dose level. The future research subject will be the effect of high radiation absorbed dose rate upon the swelling degree and mechanical strength of the above presented types of hydrogels.

References

1. Dragusin M (1989) Studies and Researches on Physics (in Romanian) 41:401

2. Dragusin M, Fiti M (1992) Romanian Journal of Physics 37:793

Progr Colloid Polym Sci (1996) 102:126–130
© Steinkopff Verlag 1996

An experimental approach to the determination of two-dimensional gel-point: A film balance study

Z. Hórvölgyi
J.H. Fendler
M. Máté
M. Zrínyi

Dr. Z. Hórvölgyi (✉) · M. Máté · M. Zrínyi
Department of Physical Chemistry
Technical University of Budapest
1521 Budapest, Hungary

J.H. Fendler
Department of Chemistry
Syracuse University
Syracuse, New York 13244-4100, USA

Abstract Two-dimensional gelation has been modelled by the lateral compression of microparticles at water(aqueous solution)–air interfaces in a film balance. The compression induced network formation manifested itself in increased surface pressure (Π) at significantly higher surface areas (A) than those observed in hexagonally close-packed particle ordering. The gel-point of a system in these experiments is defined as a surface area (or rather surface coverage) at which the particle-aggregates reach their "solid state". It can be determined by fitting a tangent to the "solid-state-part" of Π–A isotherm and extrapolating it to $\Pi = 0$.

The threshold of gelation was studied from the point of view of particle-particle interactions. The colloid and capillary interactions were influenced by the hydrophobicity of the particles and the composition of liquid phase. Computer simulations of the two-dimensional gelation were also investigated. Both the experimental and numerical results indicate decreasing gel-point (given in surface coverage) with increasing adhesion between the particles. This is in full accord with the well-known particle-particle adhesion to the specific volume of sediments (in three dimensions) relationship.

Key words Interfacial gelation – surface pressure – hydrophobicity – film balance – computer simulation

Introduction

The gel point is defined as the minimal particle concentration at which or above continuous structure, gel forms due to coagulation or flocculation [e.g., 1].

Clarifying the effect of system parameters on the gelation of interfacial solid particles, it can lead to a significant improvement of preparing thin particulate layers by a "two-dimensional" sol–gel technique for the purpose of advanced ceramics and electrooptical devices [2]. Earlier, the compressional properties of monoparticulate layers were studied at water–air interfaces from the point of view

of particle surface hydrophobicity [3–5]. Those experiments completed with some recent results are interpreted from a new aspect now. The idea is to use a film balance for the detection of contiguous particle network formation at water–air interfaces (Fig. 1).

The aim of this preliminary report is to reveal the dependence of gel-point, determined in a film balance, on the particle–particle adhesion at water (or aqueous electrolyte solution)-air interfaces. For comparison, the results of computer simulations of two-dimensional gelation from the point of view of particle-particle interaction energy is also presented.

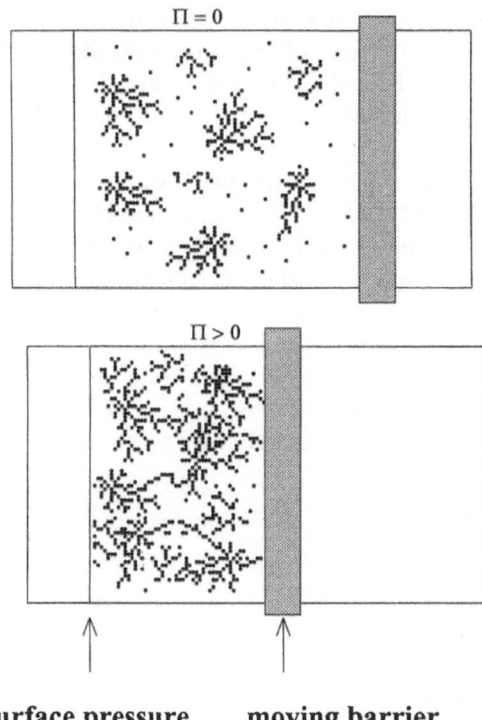

Fig. 1 Schematic picture of the "two-dimensional" system before and after the gel-point in a Langmuir trough

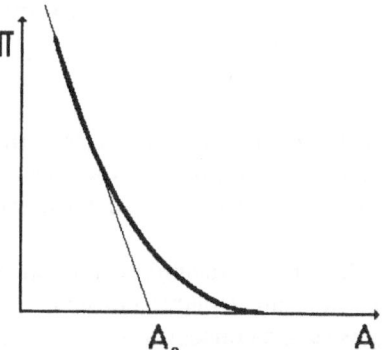

Fig. 2 Schematic drawing of a surface pressure (Π) vs. surface area (A) isotherm. The gel-point can be defined as the intersection (A_0) of the area axis and of a straight line which is fitted to "the solid state part" of isotherm

Experimental

Materials and instruments

Partially hydrophobic, spherical glass (Supelco; diameter: $75 \pm 5 \,\mu m$) and silica (Cobert Chemisorb; diameter: $3 \pm 1 \,\mu m$) particles. Both the sodium-chloride (ACS certified; Fisher) and the potassium-chloride (ACS certified; Fisher) were heated at 873 K for 1 h in order to remove any surface active contamination. For the study of the gelation a Langmuir film balance (Lauda Model P) was used. In certain cases, visual observation was accomplished by investigating the reflected light from the particles with a microscope (Olympus PM-10-M).

Experiments

Surface preparation and characterization: the surface of particles was rendered hydrophobic by silylation. The process was described elsewhere [3, 5, 6]. Water–air contact angles (Θ), measured directly on the beads, were used for the surface characterization. The details of

this optical method have also been published earlier [3, 5].

Spreading of particles at water (or aqueous electrolyte)-air interfaces in a Langmuir trough: glass beads (diameter: 75 μm) were carefully sprinkled onto the liquid surface and silica beads (diameter: 3 μm) were spread carefully from chloroform by using a microsyringe.

Surface pressure (Π) vs. surface area (A) isotherms were determined compressing the solid particles in "two-dimension" at water(or aqueous electrolyte solution)-air interfaces. *Gel-point was defined* as a surface area (A_0) which was determined by fitting a tangent to the "solid-state-part" of Π–A isotherm and extrapolating it to $\Pi = 0$ (Fig. 2). The gel-point was also given in surface coverage which means the ratio of the area covered by the particles and of the whole area of liquid-air interface bordered by the moving barrier. The effect of particle-particle adhesion on the gel-point was studied by changing the particle hydrophobicity and by introducing electrolyte into the water phase (1 M KCl in the case of glass beads and 1 M NaCl in the case of silica spheres).

Percolation model for two-dimensional gelation

Percolation theory deals with clusters which form as a result of random events. The random site percolation model [7] includes the most important feature of gelation: the appearance of an infinite network under certain conditions.

This model works as follows: let us consider a lattice and dispose particles on the sites, with no more than one particle per site. Each site is occupied at random with the probability, p_0, independent of its neighbours. If we have N sites on the lattice, then the product $N p_0$ represents the occupied sites. Thus, p_0 can be directly related to the initial particle concentration:

$$\Phi_0 = k_g p_0, \tag{1}$$

where Φ_0 represents the volume fraction of particles in the lattice (in two-dimensional case it represents the surface coverage) and k_g is a filling factor. This latter is determined by the geometry of lattice.

Particles, connected by nearest-neighbouring sites, form a cluster. Increasing p_0 results in large clusters. The percolation threshold, P_c, is such a concentration at which an infinite network appears in an infinite lattice. This threshold can be related to the gel-point. For all $p_0 \geq P_c$, one has a cluster extending from one side to the other in the lattice, whereas for all $p_0 \leq P_c$ no such "infinite" cluster (gel) appears.

In the random site percolation model the deposition is independent of the local environment, the probability of finding a particle in the neighbourhood of another one is only determined by the particle concentration. This statement cannot be true for interacting particles. To develop a more realistic model, one needs to take into account the interactions between particles.

The colloid interactions can be van der Waals attractions and double-layer repulsions, as well. Due to these interactions not every collision may be effective. If the energy barrier between approaching particles is of the order of a few kT, only a fraction of collisions leads to aggregation. In this case the efficiency of collisions must be taken into account by the aggregation probability. Thus, the probability of finding a particle as nearest-neighbour can be higher, or lower than p_0, depending on whether attractive or repulsive interactions are present. Interaction of DLVO-type can be approximated by a rectangular potential on a lattice. Infinite repulsive potential belongs to the occupied sites. The nearest-neighbouring sites for an occupied site have a potential, E, which is the interaction energy measured in kT units. The superposition of interaction energies is also considered. Thus, if a particle has q_s nearest-neighbours, then the interaction energy for this particle becomes $q_s E$. If the product $q_s E$ differs from zero, then the particle deposition is not only controlled by the average concentration, but also by the aggregation probability. The probability of finding a particle in a site can be given as follows:

$$p_s = p_0 \exp\left[-q_s E \right]. \tag{2}$$

The probability of disposition of a particle, p_s, varies from site to site together with the local coordination number, q_s. The model works as follows: a site on a square lattice is selected at random. If this site is occupied, then the procedure starts again. If the selected site is empty, then the number of the nearest neighbouring sites is determined. If p_s turns out to be smaller than a random number between 0 and 1, then the particle appears on the site. If this condition is not fulfilled, then the procedure starts again. The process continues until all of the $p_0 N$ particles are disposed.

The dependence of percolation threshold on the interaction energy was determined by the method of Vicsek and Kertész [8]. This dependence is shown in Fig. 4 and it says that the interaction energy makes its significant influence on the gelation threshold. In case of attractive interactions ($E < 0$) between particles, P_c decreases, indicating that less particles are required to form a network than for repulsive interactions. It is worth mentioning that P_c can be related to the surface coverage, Φ_c. For spherical particles on a 2D square lattice:

$$\Phi_c = (\pi/4)\, P_c. \tag{3}$$

Results and discussion

Visual observations

After spreading the particles, small and individual aggregates formed, indicating that always adhesion existed between the particles in every investigated case. The *in situ* study of the floating particles, accomplished by optical microscopy, always revealed *monoparticulate layer formation*.

Film balance results

The experimental results are collected in Table 1 where the gel-point (A_0) is given in surface area and in surface coverage, as well. As can be seen, there is clear relationship between the surface hydrophobicity and the gel-point for both the silica and glass samples. The higher the hydrophobicity of the particles the lower the surface coverage (or the higher the surface area) at which contiguous particulate layer forms. Solving electrolyte into the water phase also leads to a lower surface coverage value in the case of lower hydrophobic silica and glass particles (also see Fig. 3).

Earlier, it was pointed out that the increased hydrophobicity of interfacial particles manifested itself in greater particle-particle adhesion [3–5, 9–10] at water–air interface. The electrolyte content of water (1 M), screening the electric double layer repulsion between the beads, also can result in (increased) adhesion at water-air interface. It should be noted that for the least hydrophobic particles (glass beads, $\Theta = 55°$) at water-air interface, adhesion is simply due to the capillary attraction [11–12]. Our previous model calculations for the silica [3] and glass spheres [10], taking into account the different spreading mechanisms of silica and glass beads, also led to the above relationship between the particle hydrophobicity and

adhesion. So, the above results can be interpreted in such a way that increased particle-particle adhesion yields lower gel-point (given in surface coverage) during the "two-dimensional" gelation.

Table 1 The experimentally determined water contact angles (Θ) and gel-points (A_0) in the investigated systems. These latter parameters are given in surface area and in surface coverage, as well

	Θ	A_0	
		Surface area (μm^2/particle)	Surface coverage
Glass beads (75 μm)	55°	4936	0.895
	72°	5437	0.812
	90°	6849	0.645
	50°	5178	0.853 (0.763)*
Silica beads (3 μm)	70°	10.8	0.657 (0.627)**
	90°	12.0	0.592 (0.582)**

The results for 1 M KCl solution* and for 1 M NaCl solutions** are given in the brackets.
The standard deviation of the film balance results were below ± 5%.

Comparison of the experimental results with those obtained by computer simulations

The computer simulations also revealed the above relationship between the gel-point and interparticle interaction energies although the $P_c(E)$ (or Φ_c) values were lower than those deduced from the film balance experiments (given in surface coverage) (Fig. 4 and Table 2). The difference between the experimental and simulated surface coverage values can be attributed to the differences of real and simulated gelation mechanisms. While the particle hydrophobicity (adhesion) has only slight effect on the

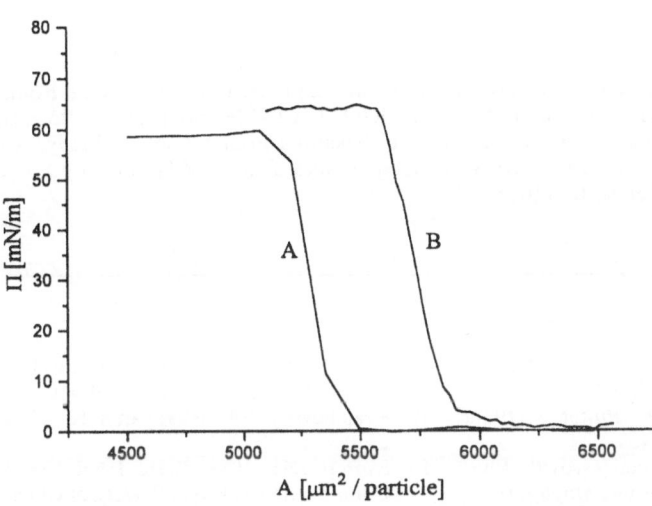

Fig. 3 Experimentally determined $\Pi–A$ isotherms of the lower hydrophobic glass spheres ($\Theta = 50°$). They were obtained for the surface of distilled water (A) and for the surface of 1 M aqueous KCl solution (B), respectively

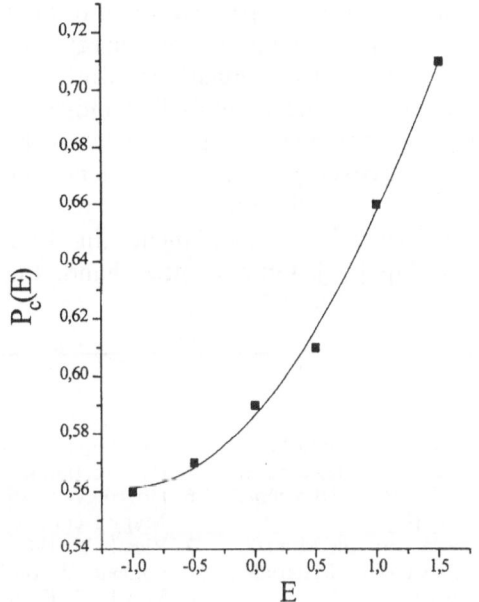

Fig. 4 The result of computer simulation: the dependence of the two-dimensional percolation threshold, P_c, on the particle-particle interaction energy (E)

Table 2 Experimentally determined and simulated gel-points given in surface coverage. The experimental values were obtained in different ways: one of them was determined as the intersection (A_0) of the area axis and of a straight line which was fitted to the "solid-state-part" of the isotherm and the other was considered as an area at Π is just greater than zero

Film balance experiments Glass spheres (diameter 75 μm) $\Theta = 55°$ ($E \approx 0$)		Surface coverage from computer simulation $E = 0$
Surface coverage (A_0)	Surface coverage (Π is just greater than zero)	
0.895	≈ 0.68	0.46

gel-point in the case of silica particles, it considerably influences the gel-point of glass beads (Table 1), indicating that the different way of spreading ("pregelation") can also influence the gel-point in the real experiments. More realistic and correct experimental value of surface coverage can be obtained at a value of Π which is just greater than zero mN/m (Table 2), but the exact determination of such values is more difficult than those determined by fitting a tangent to the "solid-state-part" of Π–A isotherm.

Conclusion

A film balance method was suggested for the determination of the (two-dimensional) gel-point of particles floating at liquid-air interface. The continuous interfacial network formation of particles (gel-point) manifested itself in measurable (greater than zero) surface pressure in a Langmuir trough. The adhesion dependence of gel-point was studied by the above method changing the hydrophobicity of particles or the electrolyte content of water phase. For comparison, the two-dimensional percolation threshold dependence on particle-particle interaction energy was also investigated by computer simulations. Both the experimental results and computer simulations indicated that the higher the particle-particle adhesion the lower the gel-point given in surface coverage. The observed tendency is in full accord with the well-known particle-particle adhesion to the specific volume of sediments (in three dimensions) relationship [13]. On the other hand, the

gel-point values which were obtained from film balance experiments were higher than those determined by computer simulations (given in surface coverage). This means that the simulation model should be modified in order to better describe the real gelation.

The following occurrences should be taken into consideration:

– *Spontaneous aggregation* of particles can occur during (and prior to) their compression.

– *Spontaneous restructuring* can take place [9–10] during the compression in the case of larger particles (diameter: 75 μm). This process can depend on the rigidity of clusters and on the capillary attraction between the cluster segments. While the cluster rigidity is under the control of particle-particle adhesion, the capillary effect depends on the size of particles. The significance of restructuring is indicated by the relatively high surface coverage values obtained experimentally. As a result of this process, much denser particulate network can form.

– The gelation in the trough is *forced by anisotropic way* [14].

– *The clusters can freely rotate* due to random hydrodynamic effects at the interface.

Acknowledgments This work was supported by the National Foundation for Scientific Research (OTKA F4216 and T015754) /to Z.H. and M.Z./, by the National Scientific Foundation /to J.H.F./, by István Széchenyi Scholarship Foundation (to Z.H.) and by Soros Foundation (to M.M.).

References

1. Petr Munk (1989) Introduction to Macromolecular Science. John Wiley & Sons, New York, Chichester, Brisbane, Toronto, Singapore, p 107
2. Fendler JH (1994) Membrane-Mimetic Approach to Advanced Materials. Springer-Verlag, Berlin
3. Hórvölgyi Z, Németh S, Fendler JH (1993) Colloids Surfaces A: Physicochem Eng Asp 71:327
4. Hórvölgyi Z, Németh S, Fendler JH (1995) Magy Kém Foly 101(11):488
5. Hórvölgyi Z, Németh S, Fendler JH (1996) Langmuir 12(4):997
6. Hórvölgyi Z, Kiss É, Pintér J (1986) Magy Kém Foly 92:488
7. Vicsek T (1989) Fractal Growth Phenomena. World Scientific, Singapore
8. Vicsek T, Kertész J (1981) Phys Lett 81A:51
9. Hórvölgyi Z, Medveczky G, Zrínyi M (1991) Colloids Surfaces 60:79–95
10. Hórvölgyi Z, Máté M, Zrínyi M (1994) Colloids Surfaces A: Physicochem Eng Asp 84:207
11. Chan FDYC, Henry JD Jr, White LRJ (1981) Colloid Interface Sci 79:410
12. Kralchevsky PA, Nagayama K (1994) Langmuir 10:23
13. Everett DH (1988) Basic Principles of Colloid Science. Royal Society of Chemistry, p 144
14. Kolb M.: Private communication

Progr Colloid Polym Sci (1996) 102:131–137
© Steinkopff Verlag 1996

C.C. Ponta
Q.K. Tran

Consolidation of porous structures by polyacrylic acid gels

C.C. Ponta (✉)
Institute of Atomic Physics
I.F.I.N.-S6
P.O. Box MG-6
Bucharest, Romania

Abstract Cultural heritage preservation is an important social activity today. Various technical solutions have been found applicable in particular cases. This paper proposes an alternative for the consolidation of the stone and wood porous structures using in situ obtained polyacrylic acid (PAA) gels. The treated sample is a composite which conserves degraded item's shape, consolidated by an internal rigid polymeric skeleton.

A three-step procedure was followed: 1) Impregnation with an aqueous acrylic acid (AA) solution; 2) In situ irradiation polymerisation; 3) Drying. Influences of the experimental parameters are discussed and a convenient process is established.

Two stone types of French origin (Tuffeau and Vassens), dry degraded wood from 17th century, and waterlogged wood from Gallo- Roman period are considered.

The consolidated structures have been characterised by added weight, superficial hardness measurements and porosimetry. They were examined by SEM photos.

Using irradiation polymerisation, one can control the temperature and prevent cracks in the treated sample.

The materials treated in conformity with this procedure could be kept in museum conditions without risks of atmospheric water uptake.

Key words Hydrogel – cultural heritage preservation – polyacrylic acid – irradiation polymerisation – composite

Introduction

The cultural heritage preservation is now an important social activity and it uses different methods and substances. The ideal substance/method for heritage items preservation (in conformity with accepted deontological principles) [1]:

- does not change the object's shape and does not alter the object's substance
  - attains the goal with minimum intervention
  - does not degrade itself
  - is easy to use and, eventually, to remove
- is not harmful for the operator
- permits the reversibility of the process

The ideal substance (or procedure) does not exist. The procedures in use are the result of a compromise, taking into account at a maximum extent:

- deontology
- the importance of the object to be treated
- the conservation conditions after treatment
- the costs

In practice, in each particular case, someone has to make a decision searching for the most appropriate substance/method to be used. The person in charge has to choose among substances/methods available. For him, the larger portfolio, the better.

Polymers are currently used in the consolidation of degraded porous materials, especially those with low molecular weight: polyethylenglycols and polyesters [2]. In a high viscous state, they are introduced in the host structure by a slow diffusion process, hardly facilitated by temperature or vacuum. In particular cases, this process may last for years.

This contribution investigates an alternative for consolidation of stone and wood degraded structures, using in situ radiation obtained PAA gels.

A three step procedure was followed:

1) Impregnation with an aqueous acrylic acid (AA) solution
2) In situ radiation polymerisation
3) Drying

Monomers, low molecular substances, are introduced into the porous structures. Irradiation polymerisation (gel constitution), followed by solvent evaporation, set up the final composite. One can count on the following premises:

– the impregnation stage, diffusion controlled, is quickly completed and the result is a homogeneous distribution of the monomer in the porous structure
– one can control the polymerisation temperature by dose rate level and by monomer concentration
– one can control the polymer added to the degraded structure by monomer concentration
– the polyacrylic acid is a somewhat sticky substance which adheres to the pores' walls
– by adding a chain transfer additive in the solution, one can control the macromolecular weight and the degree of stickiness
– in theory, the reversibility of the consolidation process is assured by the water solubility of the polymer
– the polymer has an acceptable hygroscopicity [3, vol. I, p. 79]
– the polymer is not altered at a temperature below 150 °C [3, vol. I, p. 84]

– a glass transition temperature was not identified for polyacrylic acid salts below 150 °C [3, vol. II, p. 9]
– the acrylic acid is easily radiation polymerisable [4]

Materials and methods

Acrylic acid from Aldrich, inhibited with 0.02% hydroquinone-monomethyl ether was purified by double crystallisation. NaOH from Carlo Erba, Glycerine from Prolabo, PEG 400 Breox from B.P. Chemicals were used without further purification.

The important measurements were done with the following instruments:

– irradiation doses: Red-Perspex dosimeters from Harwell, G.B.
– viscosity: Contraves TV type
– hardness: Shore D type
– porosimetry: Coultronics, Hg type
– S.E.M.: ISI-AKASI, ABT DS-130S type
– freeze-drying: Soc. SERAIL, CS 5L lab. type and VIRTIS, Freezemobile type
– Co60 facility: NUCLEART, from CEN Grenoble-France

Two limestone types of French origin: "Tuffeau" (T) and "Vassens" (V), dry degraded wood from the 17th century, and water-logged wood from Gallo-Roman period, were treated. The characteristics of the investigated samples are:

Stone-open pores, isotropic and rigid structure; the two stone types have the same porosity factor, but present great differences of the intimate structure and chemical composition [5], Table 1; samples' dimensions: $3.5 \times 3.5 \times 7.0$ cm and $7 \times 7 \times 7$ cm.
Dry wood-quasi-cellular, anisotropic, quasi-rigid structure; degradation consists in the modification of chemical composition of the pores' wall without total altering of the skeleton; samples' dimensions: $\sim 7 \times 4 \times 3$ cm.
Water-logged wood-quasi-open, anisotropic, quasi-degraded structure; the degradation process is in the stage

Table 1 Porosimetric characteristics of the selected limestones

Stone type	Porosity factor, %			Porous surface m²/g
	Global	Micropores	Macropores	
T*	35	28	7	17
V	34	9	25	0.3

* Tuffeau stone has a high silica content, due to ancient silica skeleton micro-organisms. These spherical microcomponents are responsible for the huge porous surface.

Progr Colloid Polym Sci (1996) 102:131–137
© Steinkopff Verlag 1996

of ligno-cellulose complex destruction; the remaining lignine and cellulose cannot maintain the object's physical integrity and shape without water presence; samples' dimensions: ~4 × 3 × 1 cm.

The procedure's parameters were established working on stone samples. The selected values were applied and eventually adapted to the wood samples.

Results and discussions

Stone

Aggressiveness test

The most important reason for the degradation of stone is the dissolution of contained water-soluble salts, enhanced at a low pH.

Trying to estimate the aggressiveness of the proposed procedure, we considered the hypotheses that the danger is concentrated in the impregnation stage. During this stage, the stone is in contact with the aqueous monomer solution. After the formation of polymeric gel, and especially after drying, the mass transfer is stopped. To evaluate the aggressiveness, the stone samples were sunk in two monomer solutions for 65 h. These solutions varied in AA concentration and ionising degree (I.D.) and had different theoretical aggressiveness. After that, the monomer was extracted by diffusion in deionised water. The water was changed 8 times in 34 h. Before and after the treatment, the samples were dried (18 h/60 °C) and weighed. The results (ΔM), presented in Table 2, indicate no weight-loss with one exception. Calcium acrylate that could be formed is water soluble. The weight-gain is without real significance, being a consequence of an insufficient diffusion time.

Table 2 Results of the chemical aggression test

Solution	a		b		c	
Stone Type	T	V	T	V	T	V
+/− ΔM, %	+ 0.9	+ 0.9	+ 0.8	− 0.4	0	0

a: AA 15%; I.D. 100%; pH 7; low theoretical aggressiveness.
b: AA 30%; I.D. 50%; pH 4,5; high theoretical aggressiveness.
c: deionised water.

Consolidation procedure

To establish a convenient procedure, the parameters which could have an influence on consolidation were experimented in a large range – Table 3.

Irradiation doses over 1.5 KGy assure the conversion yield to be greater than 95% [4].

The homogeneity problem

Different tests pointed out the crucial importance of

– the impregnation's homogeneity and,
– conservation of the homogeneity during polymerisation.

In the lack of the first condition, the sample cracks, probably due to the anisotropic temperature stress. This condition can be achieved by a careful choice of the impregnation solution (its viscosity and pH) combined with the impregnation method (+/− vacuum, time) [6].

The homogeneity could disappear before or during the polymerisation stage. Either large pores could become empty due to the gravity and low viscosity, or the presence of oxygen inhibits the polymerisation at the sample's surface.

Table 3 Consolidation procedure's parameters

Procedure stage	Parameter	Experimented range
Impregnation	AA concentration	15–30%
	ionising degree	50–100%
	additives	CH_3OH, C_2H_5OH glycerine, Na^+, Ca^{+2}
	immersion time	1–7 d
	vacuum ($-10^5 N/m^2$) +	0.5–10 h +
	immersion at atm. pressure	0.5–10 h
	vacuum ($-10^5 N/m^2$) +	0.5–5 h +
	immersion at elevated pressure ($+5.10^5 N/m^2$)	0.5–5 h
Polymerisation	irradiation dose, D	1.5–15 kGy
	irradiation dose-rate, d	0.16–3.0 KGy/h
Drying	temperature, T	60–130 °C
	time, t	4–96 h

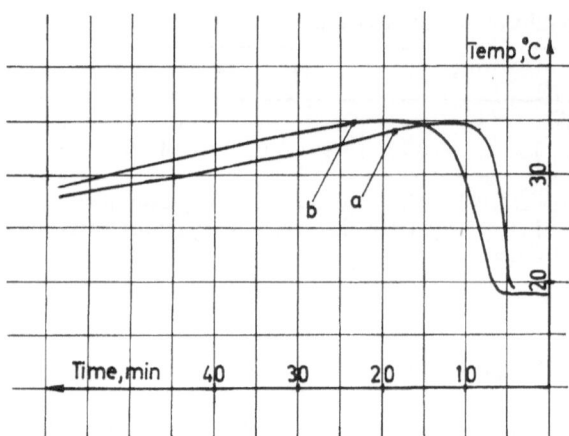

Fig. 1 Temperature evolution inside the Tuffeau stone, during polymerisation a) d = 3.6 KGy/h; b) d = 2.5 KGy/h. Impregnation solution: AA = 20%; pH = 6; glycerine 1%

To preserve the homogeneity, the liquid phase's limits has been transferred outside the sample's surface during the polymerisation stage. To attend this goal the sample impregnated with the monomer solution was sunk in a high viscous polymeric solution or coated with a soft paper soaked in that viscous solution. Due to the great difference between the viscosity of the two solutions (10–15 P versus 2–3 cP) they do not mix together. During the irradiation, "the liquid coat" is not altered. It could be removed at the end of the polymerisation stage. Because the viscous solution contains also a polyacrylate, there is only one kind of polymer in the final composite.

The polymerisation temperature

In irradiation polymerisation, one can control the temperature evolution by selecting the proper irradiation dose-rate. In the limits of the established AA concentration: 20–30%, the temperature inside the stone sample, regis-

tered with a chromel-alumel thermocouple, did not exceed 40 °C – Fig. 1.

Porosimetry

Porosimetry measurements give the possibility to evaluate the consolidation procedure and to have information about microscale structure of the composite.

Tuffeau: The consolidated measured sample has a weight-gain of 9.2%. After consolidation,

- the porosity factor diminished with ~25%
- the porous volume decreased with ~38%
- the porous surface decreased with ~49%.

These impressive values indicate the polymer covered the spongy siliceous spheres, responsible for the huge porous surface of this stone.

A sequential analysis of the porometric trial, Table 4, emphasises:

- a suppression of the pores larger than 80 μm
- a decreasing of the pores larger than 6 μm.

Together with SEM photos and hardness measurements, porosimetry brings arguments for a continuous phase in the composite.

Vassens: The consolidated measured sample has a weight-gain of 8.1%. After consolidation:

- the porosity factor diminished ~16%
- the porous volume decreased with ~18%
- the porous surface grew up with ~100%.

The growth of the porous surface is a surprise, the macroscopic image showing no damage. The sequential analysis of the porometric trial, Table 5, indicates the appearance of pores larger than 120 μm and smaller than 0.01 μm. The consolidation procedure seems to be harmful

Table 4 Tuffeau Stone's porosimetry before and after consolidation

Pores' diameter, μm	<0.02	0.02–0.5	0.5–0.1	<1.0	1.0–6.0	>6.0	6.0–80.0	>80
T-before cons., %	15.0	17.0	6.0	38.0	26.0	36.0	32.0	4.0
T-after cons., %	12.5	18.0	11.5	42.0	48.0	10.0	10.0	0.0

Table 5 Vassens Stone's porosimetry before and after consolidation

Pores diameter, μm	<0.01	0.01–0.1	0.1–1.0	<1	1–10	>10	10–40	40–120	>120
T-before cons., %	0.5	5.0	11.5	17.0	15.5	67.5	61.5	6.0	0.0
T-after cons., %	3.0	8.5	9.0	20.5	5.5	74.0	66.0	5.0	3.0

Progr Colloid Polym Sci (1996) 102:131–137
© Steinkopff Verlag 1996

Fig. 2 SEM photo of initial Tuffeau stone

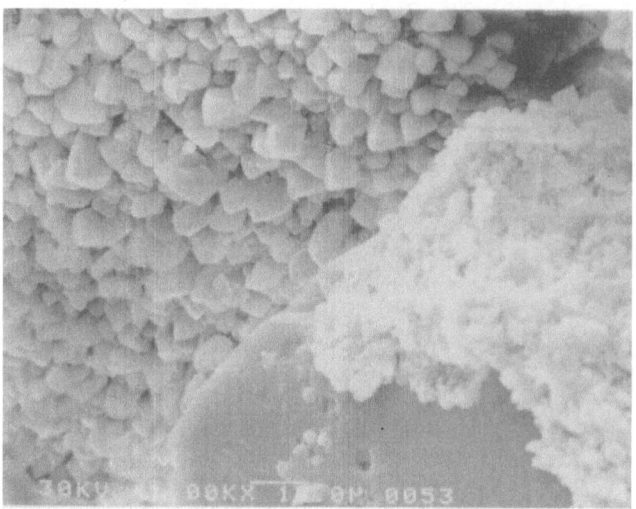

Fig. 4 SEM photo of initial Vassens stone

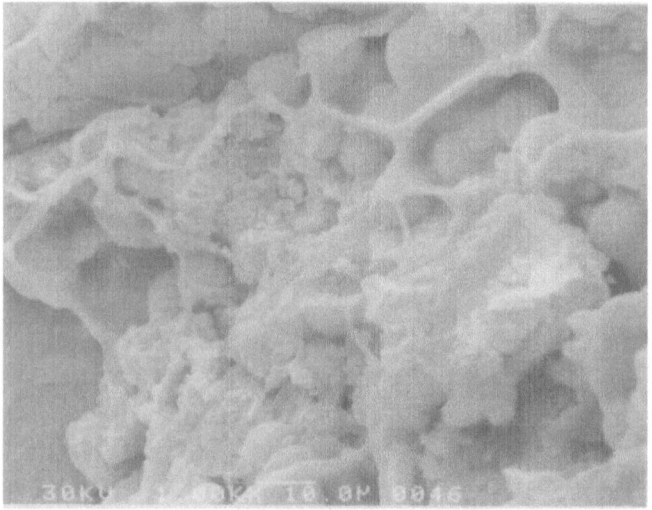

Fig. 3 SEM photo of consolidated Tuffeau stone

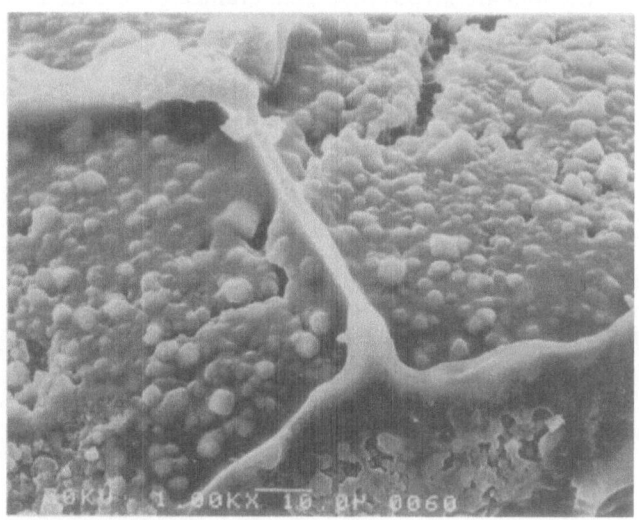

Fig. 5 SEM photo of consolidated Vassens stone

for this type of stone. SEM photos confirm the presence of microcracks. The following explanation could be advanced. Losing water, the polymeric chains relax, changing their structural conformation. Being adherent to the pore walls, they bring about microimplosions. This phenomenon is more probable in large pores.

Hardness

Hardness was measured on a fresh-cut surface and was for all stone samples close to the glass hardness, the lowest values being registered at a weight-gain under 7%. Over 9% weight-gain the hardness did not grow. A concentration higher than 20% is not useful.

Scanning Electron Microscopy

In Figs. 2 and 3, one can see SEM photos of the initial stone, respectively, the distribution of the polymer inside the Tuffeau composite. The polymer is distributed in "bridges" and coated siliceous spheres.

In Figs. 4 and 5 one can see the SEM photos of the initial Vassens, respectively, polymers distribution inside the composite structure and the presence of microcracks.

Relying on the above observations, we chose the following procedure for the stone consolidation:

Impregnation

– solution composition: AA 20%, I.D. 95% (NaOH); Glycerine 1%

pH 6; viscosity 2.8 cP

– method: vacuum ($-10^5 N/m^2/1$ h) followed by immersion at atmospheric pressure for 7 h.

Polymerisation

– irradiation dose: 1.5 KGy; irradiation dose-rate: 3 KGy/h
– liquid coating for homogeneity preservation

Drying

– temperature: 60 °C; time: 2 days

Dry degraded wood

For dry wood the higher AA concentration, the lower pH and vacuum are favourable. The retained procedure is the following:

Impregnation

– solution composition: AA 30%, I.D. 50% (NaOH); Glycerine 10%

pH 4.5; viscosity 6 cP

(glycerine acts not only as chain-transfer agent, but as softening agent as well)
– method: vacuum (-1 bar/1 h) followed by immersion at atmospheric pressure for 15 h.

Polymerisation

– irradiation doses: 2KGy; irradiation dose-rate: 3KGy/h

Drying

– temperature: 60 °C; time: 1 day

In Fig. 6, one can see a SEM photo of a dry wood-PAA composite. The polymer entirely coated the cellular wall of the consolidated wood.

Water-logged wood

The water preserves the extremely degraded structure of water-logged wood due to the fact that it is a continuous

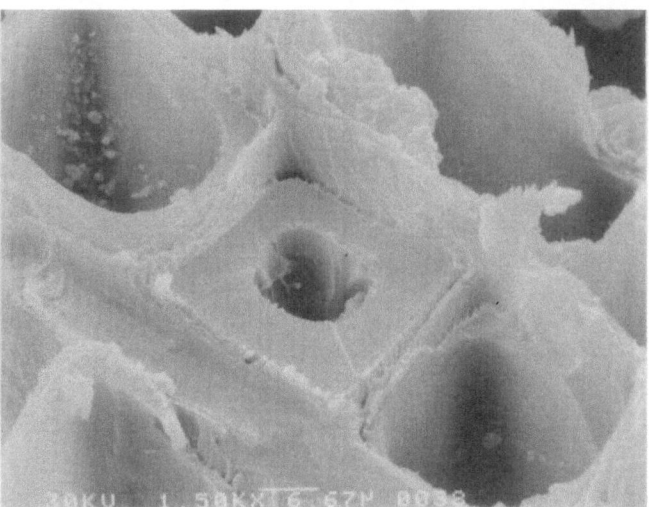

Fig. 6 SEM photo of consolidated dry wood

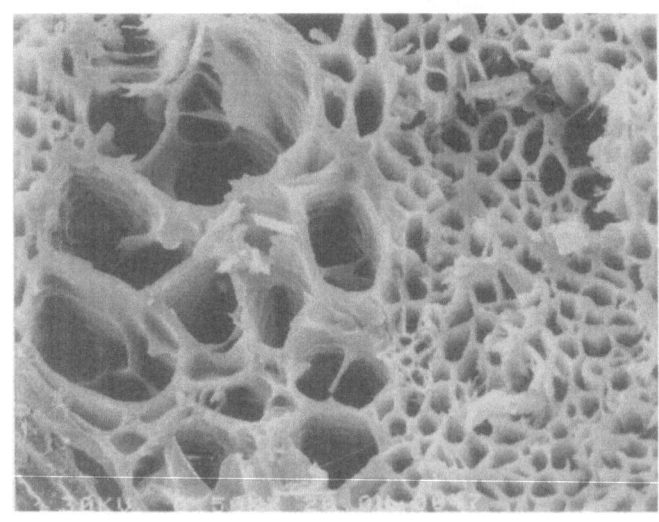

Fig. 7 SEM photo of consolidated water-logged wood

phase linked to the ligno-cellulosic complex by H bonds. The PAA could have the same properties: can form a continuous phase and is adherent.

To prevent the deformation of the sample's shape, a consequence of the polymer relaxation, water-loss must be performed by freeze-drying. Just a few experiments were done, but they clearly indicate the PAA is a crioprotector. The impregnation (without vacuum) and polymerisation stages are not critical.

SEM photo, Fig. 7, prove the consolidation of extremely degraded cell walls of the water logged wood.

Progr Colloid Polym Sci (1996) 102:131–137
© Steinkopff Verlag 1996

Conclusions

The influence of parameters

1) An acrylic acid concentration of 30%, the highest experimented limit, is convenient for dry wood; for stone, this concentration is neither favourable due to the long impregnation time, nor necessary because the same composite strength is obtained with more diluted solution. A concentration of 20% was chosen for stone.

2) A pH over 4.5 is acceptable for an impregnation time of 2–3 days. Under these conditions the monomer solution does not alter the sample's chemical composition. A strictly neutral pH, corresponding to an ionising degree of 100%, determines a greater viscosity and a longer immersion time. A pH of 6 (I.D. over 90%) for stone; a pH of 4.5 (I.D. 50%) for dry wood were chosen.

3) As chain transfer agent, Glycerine-1% was chosen. It has a reduced vapour pressure.

4) The vacuum is favourable to shortening of the impregnation time for stone. It is indispensable for dry wood due to the special structure of this material which contains captive air bubbles. A vacuum of 1 bar for 1 h was used.

5) The pressure is not necessary.

6) The samples have to be irradiated in a viscous environment, for preserving the impregnation homogeneity. A soft paper impregnated with viscous polymer solution was preferred. For dry wood this precaution is not necessary.

7) The irradiation dose of 1.5 Kgy is sufficient for a conversion yield of over 95%.

8) The maximum available dose rate of 3 Kgy/h was chosen in most experiments.

9) The drying could be performed at elevated temperature (60 °C) for stone and dry wood samples. Even temperature of 130° did not damage the samples. Waterlogged wood has to be freeze-dried.

Final remarks

1) Taking into account the great number of parameters acting on the proposed consolidation procedure, it is difficult to find an optimum procedure, in other words to justify precisely the characteristics of each stage. In these conditions a convenient procedure was identified – a procedure that succeeds in short time, with simple means and reduced costs.

2) The best results were obtained for Tuffeau type stone and dry wood.

3) Consolidation for Vassens type stone is valid from macro view point, but S.E.M. photos and porosimetry seem to indicate microcracks.

4) After some experiments with water logged wood, one can say that polyacrylic acid is a cryoprotecting substance, extremely seductive for the most problematical situations in cultural heritage preservation work.

Acknowledgements Special acknowledgements are due to M. Regis Ramiere for his useful comments and kind help.

References

1. Hovie CV (1987) Materials for Conservation. Butterworths, London, pp 5–8
2. De Agostini C, Descalle P, Tran QK (1986) Etude de la difussion dans differents calcaires de resines a plusieurs composantes, Raport NUCLEART/GR-762953, Grenoble, pp 3–4
3. Molyneaux P (1983) Water Soluble Synthetic Polymers, vol I and II, CRC Press, Boca Raton, Florida
4. Ponta CC (1992) Proceed Intl Biomed Eng Days, Istanbul: 274–278
5. Conservation of Stone (1975) Proceed Intl Symp Bologna, p 305
6. Goddard P et al (1974) J Appl Polym Sci 18:1477–1483

Progr Colloid Polym Sci (1996) 102:138–146
© Steinkopff Verlag 1996

Kinetics of sorption processes in polymer gels

V.I. Irzhak
L.I. Kuzub

V.I. Irzhak · Dr. L.I. Kuzub (✉)
Institute of Chemical Physics
Chernogolovka
Moscow distr., 142432, Russia

Abstract Kinetic features of sorption process in polymers gels with complicated supermolecular structure have been analyzed. The mechanism of parallel diffusion was suggested to describe transfer processes which consist of several partial ones carried out simultaneously in different structure elements. Each structure region is characterized by own diffusion coefficient. In this case, apparent diffusion coefficient of the transfer process as a whole is a function of time $D(t)$. The different averages of diffusion coefficient and ultimate concentration of sorbate can be obtained, using the initial rate of

process and extrapolation $D(t)_{t=0}$. These values depend on averaging type allow to relate the sorption kinetics departures from Fick's law because of heterogeneity of polymer structure or other reasons (e.g., relaxation retardation of diffusion). The experimental data were treated in framework of this model. As a result, it was shown that the approach is convenient to determine the mechanism of sorption processes in heterogeneous polymer structures.

Key words Sorption kinetics – diffusion – polymer supermolecular structure

Introduction

Polymer gels crosslinked by physical or chemical bonds possess, as a rule, inhomogeneous topological or supermolecular structure. The former is inherent to high elastic state of polymers, the later to glass state. Heterogeneity of supermolecular structure can be displayed by different manner in processes of transfer. If densities of structural elements are considerably distinguished and a part of them is inaccessible for sorbate (e.g., crystal phase in semicrystalline polymers) transfer of low molecular substances is generally suggested to be carried out in limited area of amorphous phase. This leads to decrease of effective diffusion coefficient. At the same time the sorption kinetics is described by Fick's law. In this case, structure heterogeneity, i.e., the fraction of accessible (amorphous) volume in semicrystalline polymers and winding of diffusion way,

is taken into account by correction of diffusion coefficient with numerical coefficient [1].

If fraction of forbidden diffusion ways is great, i.e., near to percolation threshold, the main peculiarity of diffusion kinetics is a dependence of diffusion coefficient on time [2,3], i.e.,

$$D = D_0 \cdot t^{-n} , \tag{1}$$

here D_0 is coefficient of proportionality, $n = 3d_f$, d_f is fractal dimension of the critical cluster.

Diffusion kinetics does not obey Fick's law, also in that case when structure heterogeneity means a few types of sorption equilibrium, e.g., which can be described by Henry's and Langmuir's isotherms. This kind of structure heterogeneity is considered in dual mode sorption model [4–6].

If the local densities of structural elements do not differ dramatically, the transfer of sorbate can be carried out simultaneously on several structural elements (diffusion channels). For example, diffusion coefficients of water into globular and interglobular areas of epoxy resins are distinguished by not more than one decade order of magnitude [1]. Obviously, in this case sorption process should be characterized by a set of transfer coefficients. However, there are practically no works analyzing kinetic features of such processes.

The present work deals with analysis of sorption kinetics features in heterogeneous polymer structures. Experimental data obtained in process of sorption of low molecular substances into amorphous and semicrystalline polymers with complicated supermolecular structure below their glass temperature have been treated to calculate the diffusion coefficients and to clarify their behavior with time.

Experimental

Subjects

Commercial fibers based on aromatic polyamides were used as sorbents. Amorphous fiber PABI (Russian trademark SVM) was manufactured by precipitation from poly(amide benzimidazole) liquid crystal solutions [7]. According to the results of x-ray diffraction [8], these fibers contain amorphous and mesomorphic regions; the latter gives meridional reflections typical of nematics. Crystalline fiber PPTA on the basis of poly-p-phenylentereph-thalamide (Russian trademark Terlon) is similar to Kevlar [9, 10] and has a 90% degree of crystallinity. Semicrystalline fiber (Russian trademark Armos) is copolymer of PABI and PPTA containing about 60% mass of PABI units [11].

Sorbates

Water vapor, liquid water and epoxy resin DGEBA was used as penetrants.

Methods of experiments

Isothermal calorimetric method was used to study sorption kinetics [12]. The experimental set-up to measure rate of heat release in process of sorption of liquid penetrants consisted of a differential isothermal calorimeter (sensitivity of the device 10^{-7} J/s, time constant 20 s) connected to a computer; experimental data were recorded each 20 s.

Fiber samples previously cleansed of impurities were placed into a glass ampoule divided into two parts by thin glass membrane. The samples were evacuated, then the ampoule was sealed off. The upper open part of the ampoule was filled with liquid sorbate and the ampoule was placed into a calorimetric cell. After thermostating of the sample the membrane was broken with a special device. Then the rate of heat release was recorded.

Although molar sorption heat of liquid penetrants was not specially determined, registrated heat release was assumed to be proportional to quantity of absorbed substance in the absence of capillary condensation.

Sorption rate and specific heat of sorption of water vapor were measured using the calorimetric set-up described earlier [13]. The set comprised two coupled isothermal calorimeters, one of which recorded the sorbate consumption by measuring the heat of evaporation, and the other one registrated the heat of interaction between the sorbate and sorbent. With the set-up it was possible to determine both rate of sorption and specific (differential) heat of sorption simultaneously as the ratio of the heat release to water vapor sorption rates at all times throughout the experiment.

The fiber samples were evacuated in a calorimetric cell at 20 °C; then, the water vapor was let in and constant pressure (1.42×10^3 Pa) was maintained by thermostating a water-filled ampoule.

Sorption of liquid water and water vapor was carried out at 20° and 30 °C. Sorption of epoxy resin was carried out at 100 °C.

Evaluation of diffusion coefficient

To determine diffusion coefficient thermokinetic curve of the sorption process was analyzed as a whole. The time interval method was proposed to solve this problem. The thermokinetic curve was divided into time intervals (≈ 500 s). Its value was negligible in comparison with whole process time ($\approx 10^4$ s). Diffusion coefficient was suggested to be constant within the time interval. Quantity of experimental points within the interval was enough to calculate the diffusion coefficient by fitting way with sufficient accuracy. So, time dependence of the diffusion coefficient could be obtained within the scope of Fick's equation.

Results and discussion

In Fig. 1 are presented the kinetic curves of sorption of water vapor and heat release. As one can see, the form of these curves is not Fickian.

Differential specific heat of sorption depends on penetration depth of sorbate (Fig. 2). Perhaps, it is an evidence of gradient heterogeneous structure of fiber skin-layer.

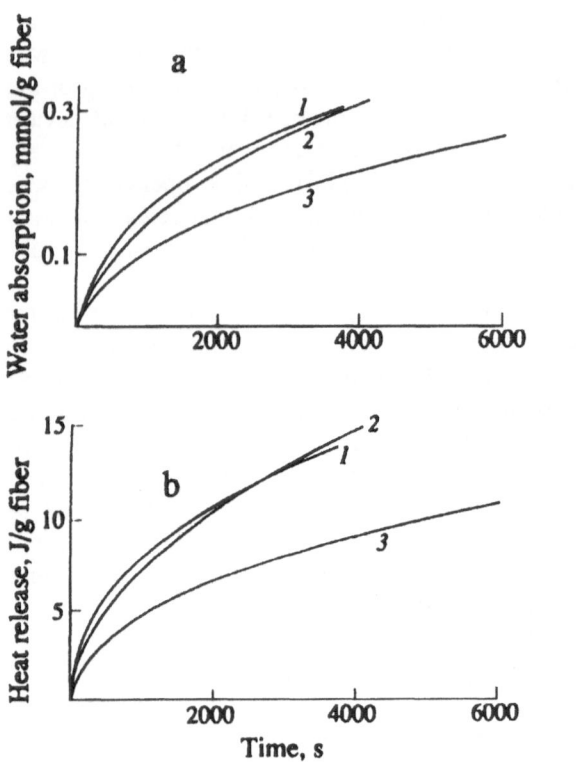

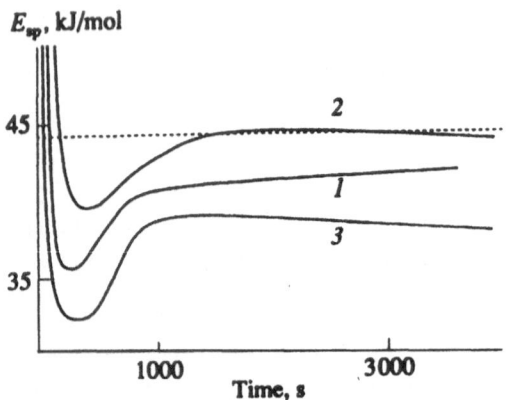

Fig. 1 Kinetic curves of water vapor sorption (a) and heat release (b). Fibers: 1-Terlon, 2-Armos, 3-PABI

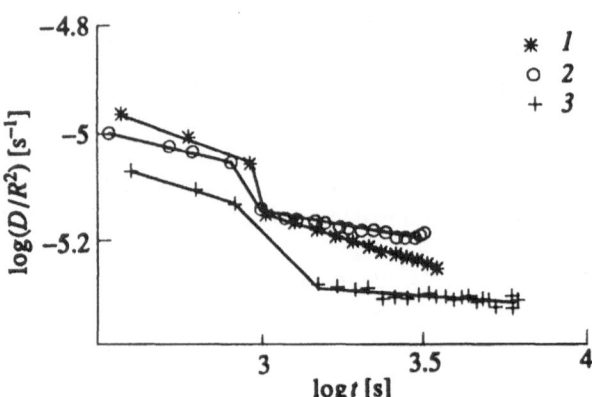

Fig. 3 The dependence of diffusion coefficient $(D/R)^2$ on time. Values D/R^2 are determined by treatment of sorption kinetic curve (Fig. 1). Fibers: 1-Terlon, 2-Armos, 3-PABI

Fig. 2 The change of specific differential heat in course of sorption of water vapor. Fibers: 1-Terlon, 2-Armos, 3-PABI. The dashed line denotes condensation heat of water vapor

Diffusion coefficients (D) determined with the interval method decrease with time (Fig. 3). There is a sharp jump on the curve $D(t)$ at time about 1000 s. This corresponds to penetration depth near to 0.1 of the fiber radius. Obviously, different structure densities of fiber skin-layer and core are responsible for this fact. Note, the decreasing of diffusion coefficient with time is observed both in fiber skin-layer and core.

The similar time dependences was obtained in studies of the processes of sorption of liquid substances: water and epoxy resin. There is only one difference: in this case the beginning of thermokinetic curve cannot be used for calculation of diffusion coefficient because of methodical features. Note, in all experiments dependence $D(t)$ was expressed by Eq. (1).

Before revealing the reasons of time dependence of diffusion coefficient, some remarks should be made.

1) The proposed interval method to analyze kinetic curves allows to determine a change of diffusion coefficient in the course of sorption process. The equation of diffusion with time-dependent diffusion coefficient is transformed to the usual Fick's equation [14], provided that τ is determined as:

$$\tau = \int_0^t D(x)\,dx \ . \tag{2}$$

A solution obtained for dummy time τ can be analyzed by comparison with the experimental curve. Because τ is time dependent a relatively narrow time interval must be considered.

If $\tau = \varphi(t)$ then

$$D(t) = d\varphi/dt + \varphi(t) \ . \tag{3}$$

That is, the dependence $D(t)$ can be determined using the function $\varphi(t)$.

2) Though the reduction of diffusion coefficient during the process does not exceed half of one order, this decrease is sufficient, as it is seen, for deviation of the curves from Fickian form. Moreover, the differences in diffusion coefficients should be taken into account because the data are obtained from the same kinetic curve, not from different ones.

Model of a parallel diffusion

As one can see from experimental data (Figs. 2, 3), investigated polymer fibers have complicated gradient structure. In this case the penetrant transfer can be considered as a set of parallel, consecutive and parallel-consecutive processes with a possibility of exchanging between diffusion channels. The most simple model of parallel diffusion is proposed here to describe diffusion in polymer body with isolated channels in the direction of a flow. The diffusion flow as a whole is represented by a sum of partial ones. Obviously, this model is suitable for polymers with developed supermolecular structure, if diffusion coefficients related to different structural regions are essentially distinguished. Clearly, this model appears to be rather artificial. Nevertheless, just this model allows to establish a compatibility between partial values of diffusion coefficients and the average ones obtained under different conditions of an experiment.

Below follows an analysis of some well known methods of evaluation of diffusion coefficients.

Steady-state flow

The diffusion equation for steady-state flow can be written as:

$$\partial c/\partial t = \sum D_i \partial^2 c_i/\partial x^2 = 0 . \qquad (4)$$

The apparent diffusion coefficient determined under these conditions is arithmetic mean, i.e.,

$$\langle D \rangle = \sum \omega_i D_i , \qquad (5)$$

where ω_i is a fraction of structure elements in which the process is characterized by diffusion coefficient D_i.

Time-lag method

By this method diffusion coefficient is calculated from time of reaching steady state (τ_{st}). The mechanism of parallel diffusion connects this value with function of distribution of diffusion coefficients by a more complicated manner:

$$\tau_{st} = \frac{I^2}{6} \cdot \frac{\sum D_i^{-1} \cdot \omega_i \cdot M_i}{\sum \omega_i \cdot M_i} \qquad (6)$$

where I is the sample size, M_i are partial ultimate sorbate concentrations.

So, experimental diffusion coefficient is determined as arithmetic mean of inverse values of partial ones:

$$\langle D \rangle^{-1} = \sum \omega_i D_i^{-1} \qquad (7)$$

Thus, the time τ_{st} is governed by the least diffusion coefficient.

These examples show that the conclusion about the effect of heterogeneity of polymer structure on diffusion process can be made only by comparison of results obtained under different conditions of carrying out the experiment.

Non-steady state

The complete information about diffusion mechanism can be obtained from analysis of kinetics of the non-steady-state sorption process. To describe kinetics of parallel diffusion direct summation of rates of partial processes should be made using the known equations for diffusion rates (w).

For a plane sample:

$$w = 8M_\infty D/I^2 \sum \exp(-D(2n+1)^2\pi^2 t/I^2) . \qquad (8)$$

For a solid cylinder:

$$w = 4M_\infty D/R^2 \sum \exp(-D/R^2 \alpha_n^2 t) , \qquad (9)$$

where 1 and R denote the plate thickness and radius of cylinder, accordingly, M_∞ is ultimate concentration of uptaken sorbate, α_n denotes the known coefficients [14].

The analysis of kinetics of the model process of parallel diffusion

The model sorption processes analyzed in this work were defined as consisting of three parallel ones characterized by partial diffusion coefficients in terms of $D/1^2$ (1 denotes a size of sample).

$$D_1/1^2: \quad 1 \times 10^{-5}\,\text{s}^{-1}; \qquad D_2/1^2 = 3 \times 10^{-5}\,\text{s}^{-1};$$

$$D_3/1^2 = 10 \times 10^{-5}\,\text{s}^{-1}.$$

Statistical weights of these partial processes were:

1) $\omega_1 = 0.6$, $\omega_2 = 0.3$, $\omega_3 = 0.1$;
2) $\omega_1 = 0.34$, $\omega_2 = 0.33$, $\omega_3 = 0.33$;
3) $\omega_1 = 0.1$, $\omega_2 = 0.3$, $\omega_3 = 0.6$.

Sorption kinetics curves calculated according to Eq. (8) for each parallel process and total one are shown in Fig. 4.

As one can see, if a process is characterized by a set of diffusion coefficients the kinetic curve describing the process as a whole does not obey Fick's law. Apparent diffusion coefficient and ultimate concentration of sorbate calculated with interval method are time dependent.

Fig. 4 Kinetic curves calculated for partial diffusion processes and whole one for plane sample: a) time dependence of rate, b) integral uptake curve. $D_i/1^2 \times 10^5$, s^{-1}: 1–1; 2–3; 3–10. ω_1: 1–0.6; 2–0.3; 3–0.11

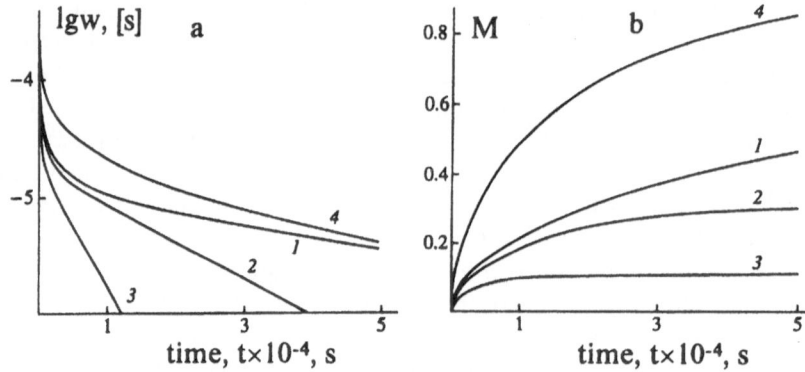

Fig. 5 The dependence of diffusion coefficients D/I^2 (a) and ultimate sorbate concentration M (b) vs. time Parameters of parallel processes: $D_1/1^2 = 1 \times 10^{-5}\,s^{-1}$; $D_2/1^2 = 1 \times 10^{-5}\,s^{-1}$; $D_3/1 = 1 \times 10^{-5}\,s^{-1}$. 1: $\omega_1 = 0.6$; $\omega_2 = 0.6$; $\omega_3 = 0.6$; 2: $\omega_1 = 0.6$; $\omega_2 = 0.6$; $\omega_3 = 0.6$; 3: $\omega_1 = 0.6$; $\omega_2 = 0.6$; $\omega_3 = 0.6$

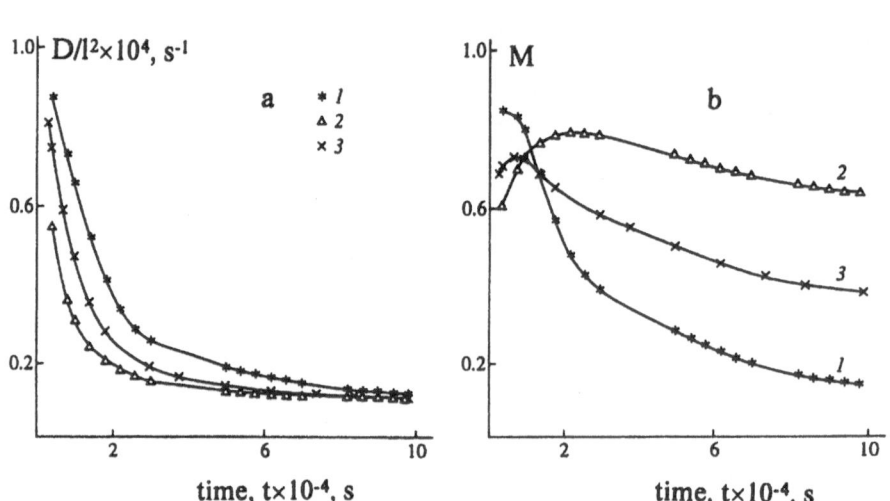

Figure 5 shows a change of $D(t)$ and $M(t)$ with time for the same set of diffusion coefficients (D_i) and different sets of statistical weights (ω_i). Both $D(t)$ and $M(t)$ are seen to be reduced with time. The curve $M(t)$ is changed non monotonically. These time dependences only reflect a complication of kinetic law of diffusion. The fact of decreasing of effective diffusion coefficient with time testifies that the diffusion processes proceed in structure regions with different diffusion coefficients.

Mathematically, dependence $D(t)$ is a result of averaging over all rates:

$$D(t) = \sum D_i \omega_i w_i / \sum \omega_i w_i . \tag{10}$$

Physical sense of $D(t)$ decreases with time in that each parallel process is terminated earlier the greater the corresponding diffusion coefficient. To each of the processes corresponds an own time of completion. At every moment average significance of diffusion coefficient depends on degree of completion of each of processes. In the beginning all of them make the contribution pursuant to their statistical weights; the contribution of slow processes increases

with time. The beginning parts of the curves (Fig. 5a) are different and depend on the set of ω_i; $D(t) \rightarrow D_{min}$ when $t \rightarrow \infty$ in any case. Thus, the function $D(t)$ comprises total information on diffusion mechanism and indicates heterogeneity of polymer structure.

The complication of sorption kinetic law does not allow to obtain function $D(t)$ in analytical form. In the present work we have to be limited by analysis of average values of kinetic parameters. An average value of diffusion coefficient is obtained as a result of extrapolation $D(t)$ to $t = 0$. In order to estimate this average the fact is used that at $t \rightarrow 0$ diffusion kinetics is well described by a decomposition on roots square of t [15]. The expression of diffusion rate for a plane sample is:

$$w = M \sqrt{D/(1^2 \pi^2 t)} + \cdots . \tag{11}$$

Taking into account Eq. (10) average value of diffusion coefficient can be obtained as:

$$D(0) = \langle D \rangle_t = \frac{\sum \omega_i M_i D_i^{3/2}}{\sum \omega_i M_i D_i^{1/2}} . \tag{12}$$

Progr Colloid Polym Sci (1996) 102:138–146
© Steinkopff Verlag 1996

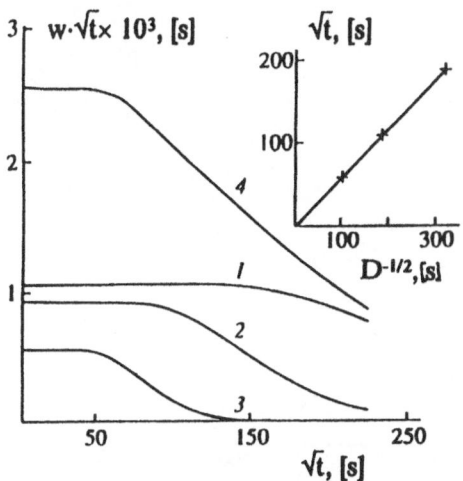

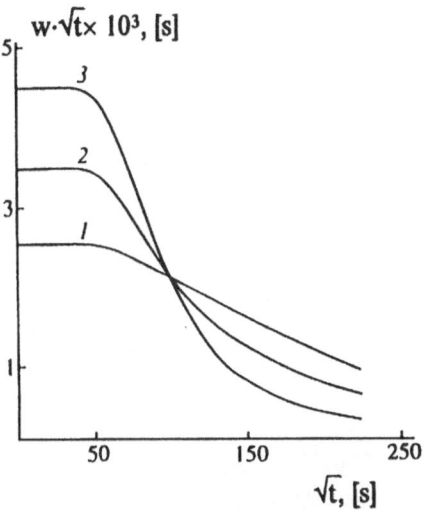

Fig. 6 Calculated kinetics curves of partial processes (curves 1–3) and of whole one (curve 4) as in Fig. 4 in coordinates of Eq. (14). The dependence of plateau extent vs. diffusion coefficient value is shown in the inset

Fig. 7 Calculated kinetic curves of parallel processes as in Fig. 5

If we compare expressions (5) and (12) obtained under both steady-state and non-steady-state regimes, it is seen that the latter expression is average value of higher order than the arithmetic mean one characterizing a steady-state flow. This distinction can be a proof that the process is carried out in a heterogeneous medium. As to ultimate concentration values (M), the dependence $M(t)$ has the same physical sense as $D(t)$, but the first one is more complicated. At the beginning it reflects (Fig. 5b) a structure of the process as a whole; then $M(t) \to \omega_{min}M_{min}$, i.e., to the value characterizing the process carried out with the least diffusion coefficient. Taking into account the decomposition (11), the expression for ultimate concentration can be obtained in the form:

$$M(0) = \langle M \rangle_t = \frac{\sum M_i^2 \omega_i D_i^{1/2}}{\sum M_i \omega_i D_i^{1/2}} . \tag{13}$$

The decomposition (11) is also useful to analyze a dependence of sorption process rate on time. Following expression (11), a product $w\sqrt{t}$ is constant at the process beginning. The beginning part of kinetic curve in coordinates $w\sqrt{t}$ vs. $\sqrt{t}$. (Fig. 6) involves a plateau with constant significance $w\sqrt{t}$. If a process is described by single diffusion coefficient ($D_i = D$, Fig. 6, curves 1–3) the plateau expansion is inversely proportional to $\sqrt{D}$, then $\sqrt{Dt} \approx 0.5$. If a process is characterized by a few values of diffusion coefficients (Fig. 6, curve 4) the plateau extension is affected by the value of maximum diffusion coefficient D_{max}. Then, a comparison D_{max} with different averages is another way to estimate structure heterogeneity. At beginning part of the kinetic curve the value $w\sqrt{t}$ (Fig. 7) is

connected with other average diffusion coefficients:

$$w\sqrt{t} \to \Sigma \omega_i M_i D_i^{1/2} = M \langle D^{1/2} \rangle , \tag{14}$$

where $M = \Sigma \omega_i M_i$.
Hence,

$$\langle D \rangle_w = \left\{ \frac{\sum \omega_i M_i D_i^{1/2}}{M} \right\}^2 . \tag{15}$$

If instead of M, its average value M_t is used, the expression (15) will take the form:

$$\langle D \rangle_w = \left\{ \frac{(\sum \omega_i M_i D_i^{1/2})^2}{\sum M_i^2 \omega_i D_i^{1/2}} \right\}^2 . \tag{16}$$

Moreover, the relation D_i/D_w can be a measure of width of diffusion coefficient distribution, i.e., to characterize a degree of structure heterogeneity.

The solid cylinder and sphere

If diffusion process is carried out in a polymer body having the form of a solid cylinder, rate of diffusion at small times and unique diffusion coefficient can be written as:

$$w \cdot \sqrt{t} = M\{2(D/\pi R^2)^{1/2} - D/R^2 \cdot \sqrt{t} - 1/2(D/\pi^3 R^2)^{3/2} t - \dots \} \tag{17}$$

The plateau on kinetic curve in coordinates $w\sqrt{t}$ vs. $\sqrt{t}$ degenerates into a point on ordinate axis at $t \to 0$ (Fig. 8). The tangent at the beginning of the kinetic curve

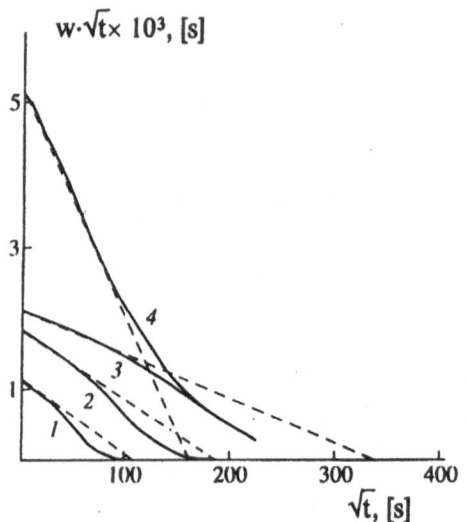

Fig. 8 Kinetics curves calculated for solid cylinder. $D_i/R^2 \times 10^5 \, \text{s}^{-1}$: (1); 3 (2); 10 (3). ω_i: 0.6 (1); 0.3 (2); 0.1 (3). Dash lines are tangents to the beginning of curves

intercepts a segment on the abscissa whose value is connected with the average significance of diffusion coefficient according to Eq. (17):

$$\sqrt{t} \cdot \langle D^{1/2} \rangle = \frac{\sum M_i \omega_i D_i}{\sum M_i \omega_i D_i^{1/2}} \cdot \sqrt{t} = \frac{2}{\sqrt{\pi}} \, . \qquad (18)$$

The length of the segment intercepted by this tangent on ordinate is defined by expression:

$$w \cdot \sqrt{t} \rightarrow \frac{2}{\sqrt{\pi}} \cdot \sum M_i \omega_i D_i^{1/2} \, . \qquad (19)$$

Both Eqs. (18) and (19) present average values of root square of diffusion coefficient. However, it should be emphasized that the averaging is carried out at different moments of distribution of D_i, and these values do not equal each other. So, this way is also suitable to estimate structure heterogeneity in cylinder polymer bodies.

As for a sphere, the beginning of the kinetic curve is described by equations similar to (18) and (19), but in this case, instead of factor 2, we have to use factor 3.

Application of the model to analysis of experimental data

The model of determination of effective diffusion coefficient in complicated supermolecular polymer structure has been applied to treat experimental data. The comparison of different average significance of diffusion coefficients, obtained from experimental dependencies $w\sqrt{t} - t$, $D(t) - t$, $M(t) - t$ (see expressions (12), (13), (18), (19)) allows to choose the most preferable sorption mechanism.

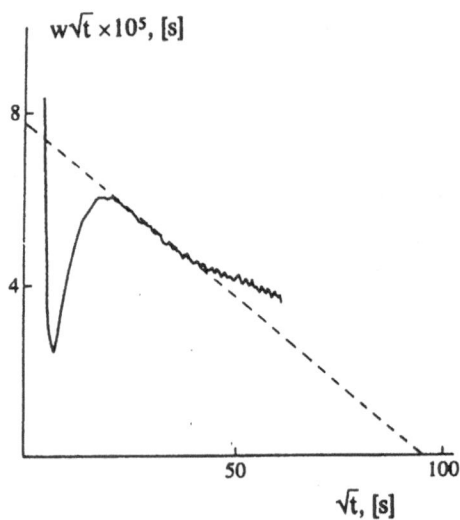

Fig. 9 Kinetic curve of water vapor sorption with Terlon in coordinates of Eq. (17). Temperature 30 °C

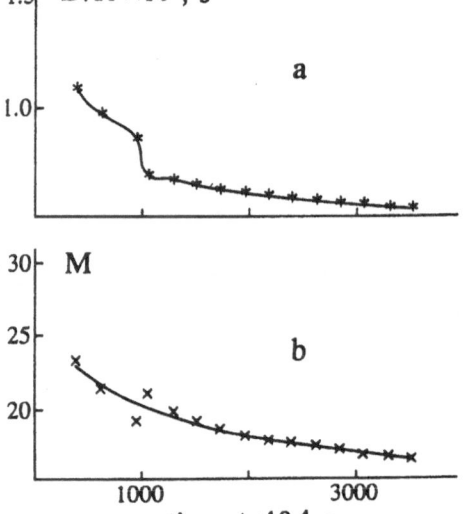

Fig. 10 The dependence of diffusion coefficient D/R^2 (a) and ultimate concentration M (b) vs. time in course of water vapor sorption with Terlon

The dependence of sorption rate on time in coordinates of Eq. (17) during sorption process of water vapor by fiber Terlon is shown in Fig. 9. As one can see, the curve has a straight segment whose extrapolation to both ordinate and abscissa axes allows to find the desired diffusion coefficients. Note, the nonmonotonic character of the curve seems to be connected with available gradient structure of fiber skin-layer. Dependencies $D(t)$ and $M(t)$ are presented in Fig. 10. Average values $\langle D \rangle_t$ and $\langle M \rangle_t$ have been found out as a result of extrapolation of these dependencies to $t = 0$. The results of calculation of these diffusion parameters obtained in this manner for a number

Table 1 The average diffusion coefficients, D/R^2, s^{-1}, calculated from kinetics data of water vapor sorption with aramide fibers

Fiber	T,°C	$\langle D \rangle_t \times 10^6$	$\langle M \rangle_t \times 10^6$	$\langle D \rangle_w \times 10^6$	$\langle D \rangle^* \times 10^6$	$\langle D \rangle_t / \langle D \rangle_w$
Terlon	20	12	25	7.9	96	1.5
	30	10	24	4.9	100	2.1
Armos	20	11	23	4.3	21	2.6
	30	10	20	4.0	10	2.5
PABI	20	9	20	3.1	90	2.9
	30	8	19	3.1	115	2.6

$\langle D \rangle_t$, $\langle M \rangle_t$, $\langle D \rangle_w$, $\langle D \rangle^*$ are calculated from Eqs. (12), (13), (16) and (18).

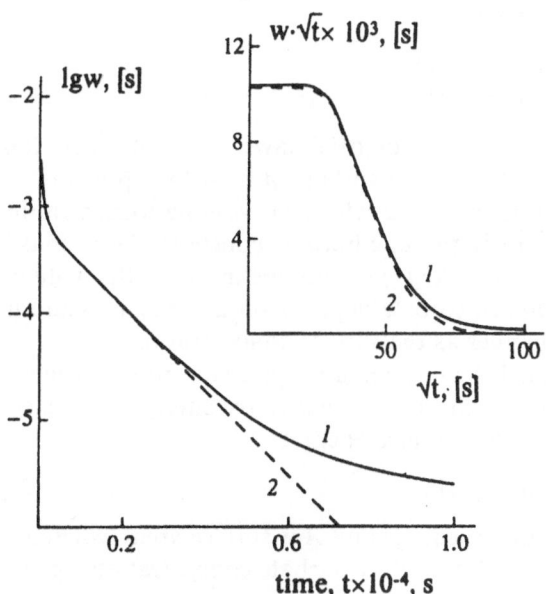

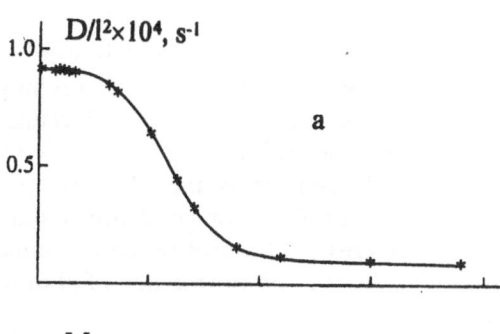

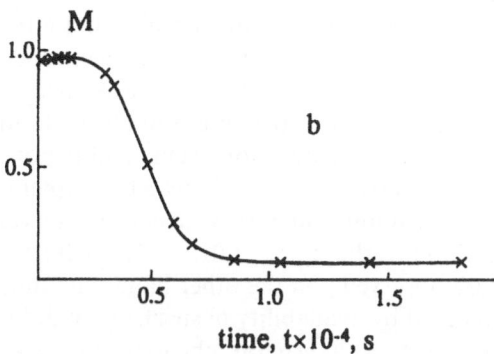

Fig. 11 Kinetic curve of water vapor sorption with PEEK (The data are taken from [15]). Dashed lines correspond to Fick's law

Fig. 12 The dependences of diffusion coefficient D/l^2 (a) and ultimate concentration M (b) on time obtained by treatment of data from [15]

of investigated polymer fibers are summarized in Table 1. The diffusion coefficients distinguished by the average type have different values. It proves that some spectrum of diffusion coefficients is really available in these polymer fiber structures.

To confirm this point of view experimental data dealing with kinetics of sorption of water vapor into amorphous PEEK [15] have been treated in accordance to the model of parallel diffusion. Diffusion parameters (w, D and M) as a function of time are shown in Figs. 11 and 12. The average diffusion coefficients calculated from these data are represented in Table 2.

In this case, kinetic curves were observed to deviate from Fick's equation. Both values D and M decrease with time. Nevertheless, values of different averages of diffusion coefficients are seen to be practically the same. This means, that the sorption process proceeds in homogeneous structure. The conclusion of the authors of work [15] that observed deviations from Fick's law are caused by relaxation retardation of the sorption process seems to be completely correct.

Table 2 The average diffusion coefficients, D/l^2, s^{-1}, calculated from sorption kinetics curves (The data of [15])

Sample	$D^0 \times 10^5$	$\langle D \rangle_t \times 10^5$	$\langle D \rangle_w \times 10^5$
1	9.3	9.2	10.0
2	9.3	9.2	11.8

The value D^0 was taken from [15], $\langle D \rangle_t$ and $\langle D \rangle_w$ was calculated according to eqs. (12) and (16).

Conclusion

The proposed method of analysis of sorption kinetics in non-steady regime of diffusion enables to obtain useful information about mechanism of sorption processes and, accordingly, about heterogeneity of polymer structure. We believe this approach is applicable to investigate different polymer gels with complicated supermolecular structure.

Different average values of kinetic parameters of sorption process (diffusion coefficient and ultimate concentration of uptaken substances) can be obtained from the beginning of kinetic curve. The advantage of such approach is obvious. It allows to characterize an initial state of polymer not changed in course of interaction with low molecular sorbate. Moreover, the direct comparison of value of ultimate concentration of uptaken substance calculated from kinetic curves with one obtained from sorption equilibrium data allows to estimate an influence of a sorbate on polymer structure.

Nevertheless, it is not clear in what measure the mechanism of parallel diffusion reflects real processes proceeding in heterogeneous structures. Additional careful investigations are needed to solve this problem. But in spite of all possible corrections and refinements the obvious conclusion is that any treatment of experimental kinetic data leads to average characteristics of a process. At the same time, type of averaging depends on the method of obtaining experimental data.

The time dependence of diffusion coefficient in the bodies with heterogeneous structure, as it was mentioned above, is considered to be connected with existence of fractal structure. Actually, such structure could be realized as the channel systems where diffusion occurs and regions where transfer is forbidden. But in this case the dependence (1) $D \sim t^n$ can be fulfilled in narrow structure interval near the percolation threshold only. Obviously, such situation is to be rare observed. On the other hand, this time dependence stipulated by availability of spectrum of diffusion coefficients must be a general phenomenon. This explanation seems to be more preferable.

One more circumstance which has to be taken into account in considering diffusion processes in inhomogeneous polymer gels is existence or absence of solid surface or, in other words, the rate of relaxation of the surface

structure. Note, the meaning "surface" related to polymer solids is implied to be the more or less expanded layer. Its thickness is much more than a mono-molecular one.

If relaxation process proceeds with sufficiently high rates, the structures of the surface and the volume will be identical. In this case equilibrium gas–polymer volume will obey Henry's law with taking into account of polymer nature using Flory–Huggins' theory. In the opposite case, when the rate of surface relaxation is rather low, one must consider the total equilibrium as based on principle of detail equilibrium:

gas-surface (I) and
surface-volume (II).

I) obeys Langmuir or BET' laws: $a = f(p)$, where a is surface concentration of sorbate, p is sorbate pressure.

II) leads to: $c = K \cdot a$, where c is volume concentration of sorbate, K is the equilibrium constant. As a consequence of it, $c = K \cdot f(p)$. This means that the volume concentration of sorbate depends on a vapor pressure in the same manner as the surface concentration.

This conclusion is readily apparent from a general thermodynamic concept. Actually, free energy of a system is connected with surface energy:

$$F = U - TS + \gamma \cdot A , \qquad (20)$$

where γ is surface energy and A is surface area. Differentiation of (20) with respect to sorbate concentration c gives:

$$\mu_v = \mu_g + A \cdot \partial\gamma/\partial c . \qquad (21)$$

Here, μ_v and μ_g are chemical potentials of sorbate in volume and gas phases, accordingly.

These considerations allow to satisfactorily describe the processes of sorption and transfer in polymer bodies without appealing to the known concept of dual mode sorption [4, 5].

References

1. Chalych AE (1987) Diffuziya v Polimernykh Sistemakh (Diffusion in Polymer Systems). Khimia, Moscow
2. Stauffer D (1985) Introduction to Percolation Theory. Taylor and Francis, London, Philadelphia
3. Sokolov IM (1986) Usp Fiz Nauk 150: 222–293
4. Petropoulos JH (1989) J Polym Sci Polym Phys Ed 27:603–620
5. Koros WJ (1993) Macromolecules 26:1493–1507
6. Petropoulos JH (1994) In: Paul DR, Yampolskii YuP (eds) Polymeric Gas Separation Membranes. CRC Press, Boca Raton, 17–81
7. Volokhina AV (1991) Khim volokna 5:7–12
8. Shuster NM, Dobrovol'skaya IP, Chereiskii ZYu, Egorov EA (1989) Vysokomol Soedin Ser B 31:348–351
9. Li L-S, Allard LF, Bigelow WC (1983) J Macromol Sci Phys 22:269–278
10. Dolb MG, Jonson DJ, Savill BP (1977) J Polym Sci Polym Phys Ed 15: 2201–2215
11. Rozhdestvenskaya TA, Tikanova LYa, Volokhina AV, Shel'din VK, Kvitko IYa, Migaev GI, Kudryavtsev GI (1989) Vysokomol Soedin Ser B 31:389–392
12. Nitkitina OV, Kuzub LI, Irzhak VI (1993) Polymer Sci Ser A 35:646–650
13. Pilyugin VV, Kritskaya DA, Ponomarev AN (1984) Vysokomol Soedin Ser B 26:907–910
14. Crank J (1957) The Mathematics of Diffusion. Oxford University, Oxford
15. Mensitieri G, Apicella A, Kenny JM, Nicolais L (1989) L Appl Polym Sci 37:381–396

Progr Colloid Polym Sci (1996) 102:147–151
© Steinkopff Verlag 1996

Polymers for waste water treatment

D. Martin
M. Dragusin
M. Radoiu
R. Moraru
A. Radu
C. Oproiu
G. Cojocaru

Dr. D. Martin (✉) · M. Dragusin
M. Radoiu · R. Moraru
A. Radu · C. Oproiu · G. Cojocaru
Institute of Atomic Physics
IFTAR-Electron Accelerator Laboratory
P.O. Box MG-6
Magurele
76900 Bucharest, Romania

Abstract Two types of anionic poly-electrolites, co-polymer of the acrylamide-acrylic acid (PA type) and co-polymer of the acrylic acid-vinyl acetate (PV type), obtained by gamma and electron beams irradiation, are presented. The experimental results concerning the typical characteristics achieved for these polyelectrolites and their applications in real waste water treatments are also presented. The influence of the chemical composition of the irradiated aqueous solutions and the radiation absorbed dose level upon the characteristics of these co-polymers are discussed. The required radiation absorbed dose levels to produce these co-polymers are rather small, from 0.4 kGy to 1 kGy for PA type and 3 kGy to 4 kGy for PV type. It is possible to obtain a very wide range of PA and PV characteristics and therefore a large area of application by controlling the chemical composition of the solution to be treated and by a suitable adjustment of the radiation absorbed dose level.

Key words Polymers – radiation – electron

Introduction

Radiation research in the field of polymeric flocculants was first developed with Co^{60} sources at the Institute of Nuclear Physics and Engineering (INPE) and, over recent years, at the Institute of Physics and Technology for Radiation Devices (IPTRD), with electron linear accelerators of 6 MeV and output power in the range of 100 to 700 W. Some polymeric materials, such as co-polymer of acrylamide-acrylic acid (PA type) and co-polymer of acrylic acid-vinyl acetate (PV type) have already been put into small commercial production with IETI-10,000 Co^{60} source. Thanks to the remarkable properties of the technologies developed for our PA and PV co-polymers, it was proved that low power-high energy linacs [1] become economically attractive for commercial production of these polymeric materials. The estimation of processing rates for a linac of 1 kW output power and 5 to 10 MeV energy is up to 2000 kg/h for the PA type.

Radiation research in the fields of anionic polyelectrolites

Preparation of PA and PV type polyelectrolites is based on polymerisation, by gamma or electron beam irradiation, of the acrylamide-sodium acrylate and acrylic acid-vinyl acetate aqueous solutions, respectively. The polymerisation of such solutions may be influenced by the following factors: chemical composition, radiation absorbed dose (D) level and radiation absorbed dose rate (d) level. The typical chemical compositions of the aqueous solutions to be irradiated are: 40% total concentration of monomers, 90% acrylamide and 10% sodium acrylate for PA type and 10% total concentration of monomers, 8% acrylic acid and 2% vinyl acetate for PV type. In addition,

D. Martin et al.
Polymers for waste water treatment

Table 1 The effect of total monomer concentration (TMC), complexing agent concentration (CAC), chain transfer agent concentration (CTAC) and radiation absorbed dose (D) upon conversion coefficient (C), molecular weight (M_w) and solubility (s = soluble, ps = partial soluble) of the irradiated acrylic acid aqueous solution

Proof	TMC %	CAC %	CTAC %	D kGy	C %	M_w	Solubility
1	30	0.025	1	1.5	72	crosslinked	ps
2	40	0.025	2	1.5	88	$2*10^6$	s
3	50	0.025	2	1.5	94	$3.5*10^6$	s
4	75	0.025	2	1.5	60	crosslinked	ps
5	100	0.025	0	3.0	10	–	–

Table 2 The effect of radiation absorbed dose (D) and chain transfer agent concentration (CTAC) upon molecular weight (M_w), conversion coefficient (C) and absorption capacity (AC) of PA co-polymer type (PA type = co-polymer of acrylamide-acrylic acid)

Proof	D (kGy)	CTAC (%)	M_w*10^{-5}	C (%)	AC (g_{water}/g_{proof})
1	3	0.00	crosslinked	≈ 100	30
2	1	0.00	crosslinked	≈ 100	42
3	0.6	0.00	crosslinked	≈ 100	150
4	3	0.10	crosslinked	≈ 100	84
5	3	0.02	partial crosslinked	96.4	partial soluble
6	3	1.00	10	96.4	soluble
7	0.6	0.05	16	98.5	soluble
8	0.6	0.10	12.5	97.6	soluble
9	0.6	0.20	8.5	98.2	soluble

we used certain agents as complexing agent (CA), chain transfer agent (CTA) and some initiators (I).

The radical reaction mechanism depends on the total monomer concentration (TMC) as well as on the water presence in the system, but the radicals originated from irradiated water have a predominant role on the radicals which come directly from the monomers irradiation. Table 1 present some characteristics of an irradiated acrylic acid aqueous solution, such as conversion coefficient (C), molecular weight and solubility (S), as a function of total monomer concentration (TMC), complexing agent concentration (CAC) and chain transfer agent concentration (CTAC). For the irradiated solution in which the water concentration decreases under 40%, the conversion coefficient decreases. In the case of the systems with very small water concentration (proof 5), the polymerisation velocity suddenly decreases, C is only 10%, although the radiation absorbed dose level is increased from 1.5 to 3 kGy and TMC is nearly 100%. The irradiated radicals presence facilitates the polymerisation process and decreases the required radiation absorbed dose (RD) level.

The radiation absorbed dose (D) level for polymerisation of acrylamide-acrylic acid aqueous solution of 40% TMC is smaller by a factor of about 3 than the level of D for polymerisation of acrylic acid aqueous solution of 40% TMC. This effect is due to the bigger acrylamide radiation reactivity ($r_{AM} = 1.6$) in comparison with acrylic acid reactivity ($r_{AA} = 0.6$). Also, because the ratio of the

monomer concentrations of acrylamide and acrylic acid is in inverse proportion to their reactivities ratio, the residual monomer concentration of acrylamide (a toxic monomer) is about 2.8 times smaller than the residual monomer concentration of acrylic acid, in the composition of acrylamide-acrylic acid co-polymer. This effect leads to the possibility to increase the limit of the total residual monomer in the finite composition of the acrylamide-acrylic acid co-polymer. The proper characteristics of the acrylamide-acrylic acid co-polymer can be well controlled by chemical composition of the irradiated solution and by a suitable adjustment of the absorbed dose level. Table 2 presents the effect of the variations in values of radiation absorbed dose (D) level and chain transfer agent concentration (CTAC) upon molecular weight (M_w), conversion coefficient (C) and absorption capacity (AC) of proof products obtained by electron beam irradiation of the acrylamide-acrylic acid aqueous solution of 40% TMC, 90% acrylamide and 10% acrylic acid. Since CTAC is below 0.2% and D is high (0.6–3 kGy), the products are crosslinked and therefore unsuitable as polymeric flocculants. Only for high CTAC, nearly 1%, the product obtained at high D is water soluble, but high values of CTAC lead to lower values of molecular weight. When level of D is nearly the proper dose (0.6 kGy), the required CTAC diminishes.

The effect of the radiation absorbed dose rate (d) level, for a given chemical composition of the acrylamide-acrylic

Table 3 The effect of radiation absorbed dose (*D*), radiation absorbed dose rate (d) and additives (initiator I and chain transfer agent CTA) concentration given in arbitrary units (a.u.) upon PA type co-polymer average molecular weight (M_w). PA type = co-polymer of acrylamide-acrylic acid

Proof	1	2	3	4	5	6	7	8	9	10
I (a.u.)	1	1	2	3	1	1	1	1	1	1
CTA (a.u.)	1	2	2	2	2	2	2	2	2	2
M_w*10^{-6}	5.0	4.6	4.0	3.7	7.4	6.3	5.0	4.0	3.2	5.0
D (kGy)	1	1	1	1	0.4	0.6	0.8	0.8	0.8	0.8
d (kGy/min)	0.8	0.8	0.8	0.8	0.8	0.8	0.8	1.6	0.4	1.2

Table 4 The effect of the *R* ratio (acrylic acid per vinyl acetate concentration) and total monomer concentration (TMC) upon conversion coefficient (*C*) and viscosity (*V*) of the PV type co-polymer (PV type = co-polymer of acrylic acid-vinyl acetate)

Proof	TMC %	R	C %	$V*10^{-5}$ cP	Physical aspect
1	8	70/80	92	5	opalescent
2	10	80/20	96	0.8	transparent
3	12	75/25	88	6	opalescent
4	15	80/20	90	2	transparent

acid aqueous solution is also very important. The polymerisation process is incomplete under irradiation at high values of d and continues after irradiation for an uncontrollable time period. Some results are given in Table 3. The proofs from 1 to 10 have 40% TMC, but different values for I and CTA concentrations. The concentrations of I, CTA and CA are given in arbitrary units in Table 3 as follows:

– CTA concentration of the proof 2 is two times higher than the proof 1;
– CTA and I concentrations of the proof 3 are two times higher than proof 1;
– CTA concentration and I concentration of the proof 4 are two and three times higher, respectively, than proof 1;
– The proofs from 5 to 10 have the same chemical composition as proof 1.

In Table 3, the conversion coefficients *C* were nearly 100% for all proofs from 1 to 10. Also, all proofs are water soluble, but the proofs from 1 to 4 were more difficult to dissolve. For the proof 8, obtained at twice absorbed dose rate level than other proofs, polymerisation process was incomplete under irradiation and continued after for an uncontrollable time period. This case is not attractive from the technological process point of view, nor is the proof 10 case (small d) which needs a long time for a polymerisation at good parameters. Proofs 5 and 6 are the best of all: their higher molecular weight leads to better flocculation properties. The molecular weight M_w given in Table 3 is on average, determined from Mark–Houwink–Sakurada

Table 5 Absorbed dose (*D*) effect upon conversion coefficient (*C*) and viscosity (*V*) of the PV type co-polymer (PV type = co-polymer of acrylic acid-vinyl acetate)

Proof	D kGy	C %	$V*10^{-4}$ cP
1	2	86	2
2	2.5	90	4
3	3	92	5.2
4	3.5	96	6.3

relation [2], on the basis of measured proofs intrinsic viscosity.

The PA co-polymer type exhibits a very low toxicity but a total conversion of acrylamide (a toxic monomer) cannot be obtained by any polymerisation method. For that reason, research has been carried out in IAP to obtain the PV co-polymer type which is based on monomers without toxicity. The experimental results concerning PV co-polymer type are given in Tables 4, 5. Table 4 presents the effect of the total monomer concentration (TMC) and the ratio *R* (acrylic acid concentration per vinyl acetate concentration), for a constant absorbed dose *D*, upon conversion coefficient *C* and viscosity *V* of the PV copolymer type. The analysis of the results given in Table 4 indicates that proof 2 is suitable as polymeric flocculant. According to this first conclusion, we finally used in our experiments a solution with 10% TMC and R = 80/20. Table 5 gives for this solution the effect of the absorbed dose *D* upon *C* and *V* of the finite products. The proof-product with the best characteristics was obtained for *D* of 3.5 kGy.

150

D. Martin et al.
Polymers for waste water treatment

Table 6 Typical characteristics of PA and PV type co-polymers (PA type = co-polymer of acrylamide-acrylic acid and PV type = co-polymer of acrylic acid-vinyl acetate)

Type	PA	PV
Physical aspect	transparent granular gel	transparent viscous solution
Polymer content	40% $\pm$ 2%	10% $\pm$ 1%
Water content	up to 50%	up to 90%
Anionic charge	0–30%	80% $\pm$ 1%
Molecular weight	$(3-10) * 10^6$	$(1-2) * 10^6$
pH of 1% solution	$7 \pm 1\%$	$4 \pm 1\%$
Weight/unit volume	$(1.16 \pm 0.5\%)$ g/cm^3	$(1.06 \pm 0.5\%)$ g/cm^3
Feed solution	0.05%–0.1%	0.5%–1%

Table 7 Results obtained by testing PA and PV co-polymers with waste waters from a sanitary objects factory and food industry. The treated waters were: type 1 – water from sanitary object icing steady; type 2–water from mass preparation steady; type 3 – water from bread yeast preparation; type 4 – water from beer pasteurisation; type 5 – water from slaughter house (PA type = co-polymer of acrylamide-acrylic acid and PV type = co-polymer of acrylic acid-vinyl acetate)

Water type	1	2	3	4	5
Solid substance concentration	19%	3%	0.04%	0.07%	0.11%
pH	6.5	6	7	5.5	5
Waste water light transmittance (500 nm)	0%	0%	88%	49%	0%
Treated water light transmittance (500 nm)	94%	91%	96%	87%	93%
Distilled water light transmittance (500 nm)	98%	98%	98%	98%	98%

Table 8 Standard specifications for initial waste water proofs and final water proof treated with PA and PV co-polymer types (water proof 3 – from bread yeast preparation; water proof 4 – from beer pasteurisation; water proof 5 – from slaughter house)

Water type	3 waste/treated	4 waste/treated	5 waste/treated
pH	6.4/7	5.8/7	6.8/8
mg KMnO$_4$/l	613/51	1070/68.5	1580/82.2
mg O$_2$/l	32.4/30	62.6/42.3	217.6/45.2

PA co-polymer type is used for industrial water treatment and PV co-polymer type for surface water treatment.

The typical characteristics obtained for PA and PV type polymeric flocculants are given in Table 6.

The main results obtained by testing PA and PV polymeric flocculants with waste water from a sanitary objects factory and food industry are given in Table 7. The PA co-polymer type is used in aqueous solutions with a concentration of 0.05%–0.1% and PV co-polymer type with a concentration of 0.5%–1%. PA co-polymer could be very efficiently used for hydrometallurgy without adding other inorganic flocculants (aluminium sulphate or ferric chloride). For instance, at a kaolin (porcelain earth) mining plant which produces residual water with density of 1010 kg/m^3–1033 kg/m^3 (80% kaolin particles size having 2 μm), we used about 100 g–120 g of PA co-polymer per 1000 kg of dry solid substance. In this case, final sediment density was 1100 kg/m^3–1200 kg/m^3 and the sediment thickness was 190 mm–210 mm for the first minute. In many cases the PA and PV co-polymer types are used together with inorganic flocculants but their consumption may be reduced from 50% to 75%. For example, residual water of type 3 (from bread yeast preparation process) was treated per 100 ml with 0.04 ml PV co-polymer solution of 1% and 0.08 ml FeCl$_3$ solution of 10%. Table 8 gives some standard specifications for initial waste water proofs and final treated water proofs of type 3 (bread yeast preparation process), type 4 (beer pasteurisation process) and type 5 (from a slaughter house).

Conclusions

The properties of PA and PV co-polymer types, obtained by gamma and electron beam technologies, may be well controlled by radiation absorbed dose level, radiation absorbed dose rate level and chemical composition of the aqueous solutions to be irradiated. The required radiation

Progr Colloid Polym Sci (1996) 102:147–151
© Steinkopff Verlag 1996

absorbed dose levels to produce these co-polymers are rather small, from 0.4 kGy to 1 kGy for PA type and from 3 kGy to 4 kGy for PV type. These ranges depend on the solution chemical composition and absorbed dose rate level. The radiation absorbed dose rate level is very important: at high absorbed radiation dose rate level, the results could be poor. Our future research subject will be the effect of high absorbed radiation dose rate upon radiochemical efficiency and polymerisation velocity for PA and PV co-polymer types.

References

1. Martin D, Fiti M, Radu A, Dragusin M, Cojocaru G, Margaritescu A, Indreas I (1995) Radiat Phys Chem 435:615

2. McCarty KJ (1987) Appl Polym Sci 33:401

Progr Colloid Polym Sci (1996) 152
© Steinkopff Verlag 1996